Gene Cloning

The Mechanics of DNA Manipulation

Outline Studies in Biology

General Editors

Gene Cloning

The Mechanics of DNA Manipulation

David M. Glover

Cancer Research Campaign
Eukaryotic Molecular Genetics Group
Department of Biochemistry
Imperial College of Science and Technology
London

London New York
CHAPMAN AND HALL

First published 1984 by
Chapman and Hall Ltd
11 New Fetter Lane, London EC4P 4EE
Published in the USA by
Chapman and Hall
29 West 35th Street, New York NY 10001
Reprinted 1985, 1986

Printed in Great Britain at the
University Press, Cambridge

ISBN 0 412 26600 8 (HB)
ISBN 0 412 25430 1 (PB)

British Library Cataloguing in Publication Data

Glover, David M.
Gene cloning.—(Outline studies in biology)
1. Molecular Cloning 2. Recombinant DNA
I. Title II. Series
574.87′3282 QH442.2

ISBN 0-412-26600-8
ISBN 0-412-25430-1 Pbk

Library of Congress Cataloging in Publication Data

Glover, David M.
Gene cloning.

Rev. ed. of: Genetic engineering, cloning DNA. 2nd ed. 1980.
Bibliography: p.
Includes index.
1. Molecular cloning. 2. Recombinant DNA. I. Glover, David M. Genetic engineering, cloning DNA. II. Title.
QH442.2.G56 1984 574.87′3282 84-9575
ISBN 0-412-26600-8
ISBN 0-412-25430-1 (pbk.)

Contents

Preface

This book was originally conceived in the form of a second edition of a volume published in 1980 in Chapman and Hall's 'Outline Studies in Biology' series and entitled *Genetic Engineering – Cloning DNA*. It very rapidly became apparent that with the impact of recombinant DNA techniques being felt in so many areas of biology, it was going to be difficult if not impossible to keep the book within the space confines of these little monographs. The stays were therefore loosened and the book expanded comfortably to its present size. I hope that this extra space has allowed me to clarify sections of the text that were 'heavy going' in the earlier version. The extra space has certainly allowed me to cover topics that were not mentioned at all in the earlier book. These are primarily to be found in Chapters 7 and 8, which cover the rapid advances that have been recently made in the use of plant and animal cells as hosts for recombinant DNA molecules. The development of other vectors has certainly not stood still over the past four years. This has necessitated a thorough revision of Chapters 3 and 4, which deal with bacteriophage and bacterial plasmid vectors. Numerous techniques for *in vitro* mutagenesis have now been tried and tested allowing me to give comprehensive coverage of this area in Chapter 2 along with the biochemical techniques used to construct recombinant DNA molecules. Readers with some background knowledge of the approaches to gene cloning will be able to go straight to a part of the book in which they are specifically interested. On the other hand, undergraduate readers with little background knowledge should benefit from studying the first chapter in which the basic principles of gene cloning are expounded.

The topics covered by the book are taught in a four week course, 'Genetic Engineering', given for final year undergraduates studying Molecular Biology at Imperial College. The book is intended for use in such advanced undergraduate study and also for graduate school coursework. I hope that it will be read as well by research workers wanting a general overview of the subject. When I first reviewed the subject of Gene Cloning in an article written in late 1975, I cited 38 papers which covered the whole field. Nowadays, it is difficult to find any paper dealing with the molecular biology of Nucleic Acids that does not use DNA cloning technology. Consequently it has been difficult to be selective in providing references. I hope that students will not feel overburdened by the references that I have given and that researchers will not feel neglected if their particular area has not been covered.

Preparation of the manuscript has taken considerable time and I am grateful to my research students and other colleagues for their tolerance of my involvement in the project. Finally I am grateful to Jean Beggs, Conrad Lichtenstein and Peter Rigby of Imperial College and Bill Brammar of Leicester University for their helpful comments on the manuscript at various stages of its evolution. Finally, I thank Martin Eilers, Manfred Frasch, Herbert Jäckle and Harry Saumweber of the Max Planck Institut für Entwicklungsbiologie in Tübingen for their help in reading the proofs.

Imperial College David M. Glover

Cover picture

The photographs on the cover illustrate the procedure of colony hybridization explained in Section 1.2 (see also Fig. 1.1). The upper photograph shows a 'master' plate on which there are colonies of *Escherichia coli* containing recombinant plasmids spotted out to form a grid. The master plate has been used to 'print' out a number of replicas onto nitrocellulose filters. These have been incubated on nutrient agar allowing new colonies to grow in identical positions to those on the master plate. The nitrocellulose filters are later removed, and the DNA in the colonies denatured and immobilized on the filter using procedures described in Section 1.2. The filters are incubated with a radiolabeled nucleic acid probe which hybridizes to any complementary nucleic acid sequences in the colonies. The unhybridized radioactivity is washed away and the position of hybridization revealed by autoradiography. The lower photograph shows such an autoradiograph in which the probe has hybridized to DNA from two of the colonies. This particular experiment was carried out by Sarah Millar.

1 The principles of cloning DNA

Past progress in understanding the molecular biology of prokaryotic gene expression has relied heavily upon studies involving bacteriophage and bacterial plasmids. Of the bacteriophage themselves, the *E. coli* phage λ is perhaps the best characterized. The interaction of phage λ with the host cell is a particularly fruitful area of study, as here are a set of genes which can either direct cell lysis or become stably associated with the host chromosome in lysogeny. In the production of infectious phage from lysogens, the excision of the phage genome from the *E. coli* chromosome is usually precise. Occasionally, however, the excision is imperfect and results in a λ phage transductant which carries a segment of the bacterial chromosome that was adjacent to the phage attachment site. Such specialized transducing phage have been invaluable, providing the means to assay for specific messenger RNAs by nucleic acid hybridization or enabling the production of large amounts of particular gene products. Research on the bacterial plasmids has had a similar history. The discovery and rationalization of the mechanism whereby F factors promote bacterial conjugation was central to the development of *E. coli* genetics. Just as the imperfect excisions of phage λ from its lysogenic state can result in a circular phage genome carrying a segment of bacterial DNA, so the imperfect excision of an F plasmid from an Hfr strain results in an F′ plasmid which also carries a segment of bacterial DNA. Such F′ plasmids have been invaluable 'vectors' for carrying specific genes from one *E. coli* strain to another and have perhaps been most useful in the construction of merodiploid strains which have enabled the elucidation of the control circuits of many bacterial operons.

The principles of genetic engineering which are described in this book are analogous to these 'natural' events, but they overcome the limitation of an absolute dependence upon the *in vivo* recombinational mechanisms of the *E. coli* cell. The techniques for recombining DNA *in vitro* will be described in detail in Chapter 2. It is possible to insert DNA from any organism into a plasmid or viral replicon to form a chimaeric molecule. The host for this molecule can be a prokaryotic or a eukaryotic cell depending upon the replication origin present in the vector. The methods that were originally described for constructing recombinant DNA molecules *in vitro* were not straightforward and involved many enzymatic steps [1, 2] (see Section 2.1.3). Subsequently it was realized that certain restriction endonucleases generate cohesive ends when they cleave DNA. DNA molecules possessing these cohesive ends can be easily rejoined using DNA ligase [3] (see Section 2.1.2). The procedures for joining DNA molecules *in vitro* thereby became very much simpler and within the scope of many laboratories.

1.1 The debate on the safety of work with recombinant DNA

The techniques mentioned in the previous section, were originally applied to the joining of DNA from the tumour virus SV40 with bacterial plasmid DNA. These recombinants were not introduced into *E. coli* because a great deal of concern was voiced within the scientific community about the hypothetical hazards of introducing the DNA of a tumour virus into a bacterium capable of growing within the mammalian intestinal tract. Subsequently this concern was to spread to other experiments: perhaps there were genes within the mouse genome, for example, that could be dangerous when introduced into and expressed within *E. coli*. A meeting of scientists was called at Asilomar in California in 1975 to discuss these issues. An early assessment of the problems as they were then perceived can be found in the summary statement of the conference [4]. In the absence of evidence viewpoints became sharply polarized, and the whole issue provoked much debate and many reports from governmental and scientific bodies [5, 6]. At that time the publications from such bodies considerably outweighed scientific reports of experiments using recombinant DNA techniques. Fortunately that situation did not persist for long and publications which utilize recombinant DNA techniques now dominate the literature in the field of molecular biology. The main fear was that *E. coli* carrying a potentially dangerous cloned gene could accidentally escape the laboratory and successfully colonize the intestinal tracts of laboratory workers and precipitate some disastrous pandemic. The other point of view was that prokaryotic organisms within nature were frequently in contact with and must therefore take up eukaryotic DNA from decaying plant and animal matter. It is then likely, given the enormity of the earth's population of microorganisms, that recombination analogous to the type that can now be carried out *in vitro* has already had the chance to occur and that the resulting 'recombinant organisms' have no selective advantage.

The concern about the biological safety of these experiments stimulated work on the development of biologically 'safe' host–vector systems. These utilized vectors that would only grow in certain strains of *E. coli* that could not survive outside the laboratory. Some of these developments will be encountered in Chapters 3 and 4, in which the bacteriophage and plasmid vectors of *E. coli* are described. The standard laboratory strains of *E. coli* have now been cultured for several decades and are very much enfeebled, and have low survival in the human gut. In virtually all cloning experiments, the recipient *E. coli* strain is deficient in its endogenous restriction system (see Section 2.1.1). This ensures that foreign DNA will not be degraded when it is introduced into the bacterium during cloning. These strains are consequently highly susceptible to bacteriophage infections. Furthermore the strains are often also deficient in one of the recombination pathways, usually *rec*A. This makes the bacteria deficient in their ability to promote homologous recombination, an essential factor if, for example, one is propagating tandemly arranged genes from a higher eukaryote. *Rec*A$^-$ strains are, however, extremely sensitive to ultraviolet light. It is possible to build in many more safeguards to ensure that a bacterium containing a potentially hazardous DNA molecule has a very low chance of survival outside the laboratory environment. One such strain was developed after the Asilomar conference and was available in the year in which the United States celebrated 200 years of independence and was appropriately christened χ1776 [7]. This strain contains mutations which make its growth absolutely dependent upon the presence of diaminopimelic acid and thymidine in the culture medium. Diaminopimelic acid (DAP) is an intermediate in the biosynthesis of lysine and is not found within the mammalian intestinal tract. The strain cannot therefore survive if it is ingested. It also shows increased sensitivity to ultraviolet irradiation as a result of removal of one of the genes responsible for DNA repair, *uvr*B, and it is very sensitive to both bile salts and ionic detergents. The strain is consequently difficult to handle in the laboratory unless scrupulously clean, detergent-free glassware is used. It is, however, possible to transform the strain with efficiencies comparable to those obtained with other commonly used recipient strains. Nowadays χ1776 is rarely used, since many of the hypothetical hazards of recombinant DNA work are now viewed from a more rational perspective and in

1979 many of the guidelines for work with recombinant DNA were relaxed. There is now some ten years experience of working with eukaryotic DNA cloned in bacterial cells. In the process an enormous amount has been learned about the eukaryotic genome which suggests that many of the early fears were ill-founded. Largely as a result of the exploitation of the recombinant DNA techniques, we now know that most eukaryotic genes have a pattern of chromosomal organization that precludes their expression in prokaryotic cells. In order to get expression of eukaryotic genes in *E. coli* it is usually necessary to clone either a DNA complement of the mRNA of interest or a chemically synthesized gene that lacks the intervening sequences commonly found in eukaryotic genes. Furthermore these must be linked to the correct prokaryotic signals for the initiation of transcription and translation (see Chapter 5).

1.2 General principles of cloning

The vectors that have been specifically designed to carry foreign DNA sequences have extended the traditional genetic approach developed with transducing phages and F plasmids. The techniques for cloning DNA were worked out using *E. coli* as the host organism. In Chapters 3 and 4 the phage and plasmid cloning vectors of *E. coli* will be described in detail. A vector should have its own replicon and thereby be capable of autonomous replication in the host cell; it should carry one or more selectable marker functions, in order to permit the recognition of cells carrying the parental form of the vector or a recombinant between the vector and foreign DNA sequences; finally, it should have sites for a variety of restriction endonucleases which cleave the molecule once or, at the most, twice in non-essential regions so that foreign DNA may be inserted into the vector or used to replace a segment of the vector. For the purposes of this introductory discussion, we will consider the general experimental approach for cloning DNA into the circular DNA of a plasmid vector or into the linear DNA of a vector derived from bacteriophage λ. This approach is illustrated for a plasmid vector in Fig. 1.1. The variations on this scheme applicable to cloning in λ vectors are described in outline later in this section and in detail in Chapter 3. The circular plasmid DNA must first be cleaved, preferably at a unique site to generate a linear molecule. One of the several techniques described in Chapter 2 is then used to join this linear vector molecule to the foreign segments of DNA that are to be cloned. A number of products can result from this reaction. In terms of the molecules that are subsequently capable of replicating in *E. coli* one has to consider both the formation of circular recombinant DNA molecules and the possibility that a significant proportion of circular

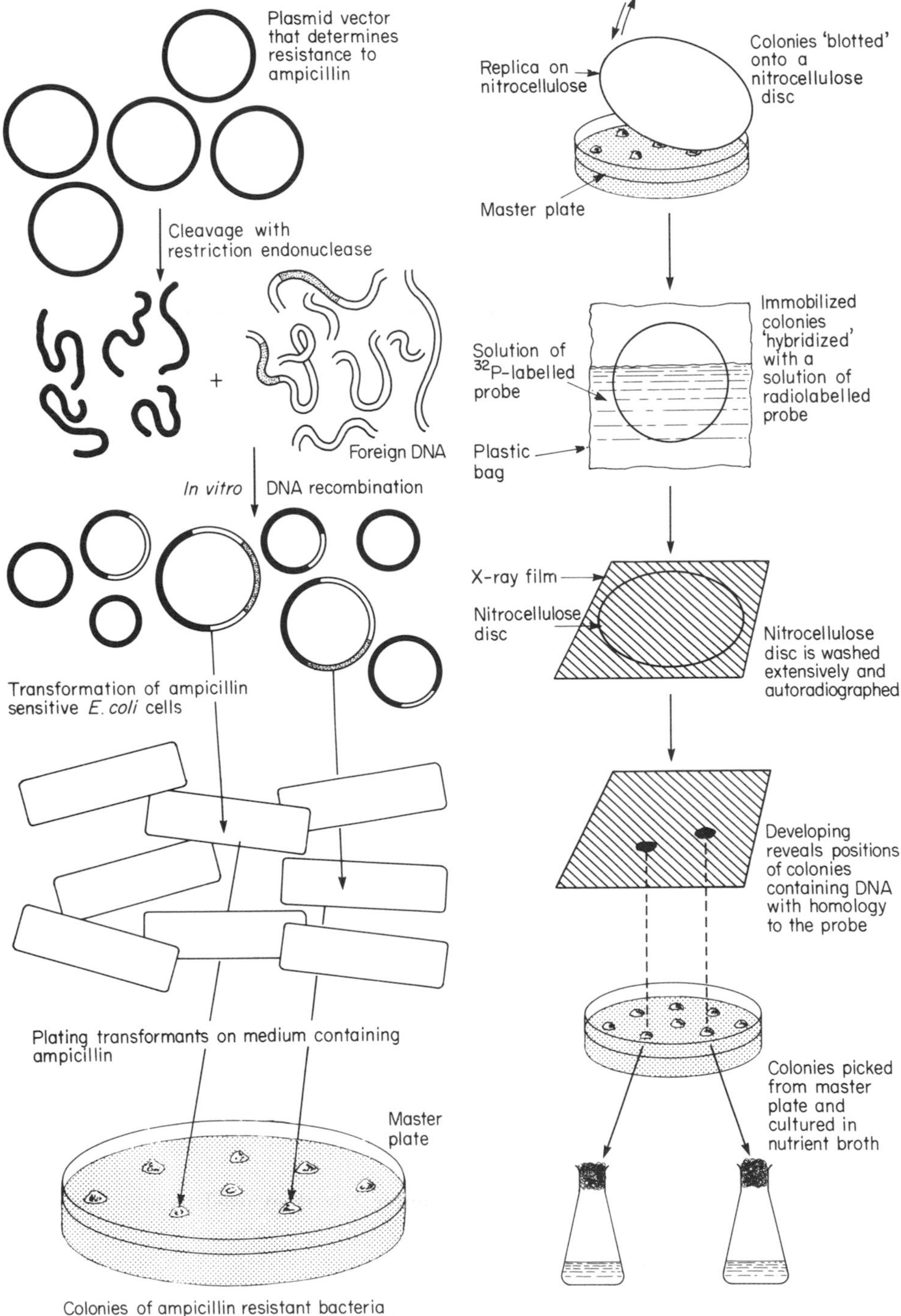

Figure 1.1 A general scheme for cloning DNA in a plasmid vector.

vector molecules can be regenerated. (Linear plasmids, recombinant or parental, transform *E. coli* at least an order of magnitude less efficiently than do circular molecules.) The products of the reaction are thus a highly heterogeneous mixture of recombinant molecules together with parental plasmids. This mixture is usually introduced directly into *E. coli* by a variation on the transformation procedure originally developed by Mandel and Higa [8]. In this procedure DNA is incubated with *E. coli* cells that have been made permeable by treatment with Ca^{2+} ions. The procedure was first used as a means of introducing λ DNA directly into *E. coli* and so acquired the term 'transfection' (a hybrid of transformation and infection). This term is still generally applied to the procedure. The efficiency of this process was such that about 10^5 transformants could be generated per μg of a covalently closed circular plasmid such as pBR322. The experimental parameters for transfection have now been more rigorously examined [9] and it is possible to achieve efficiencies of over 10^8 transformants per μg using the protocols that have emerged. DNA cloned in the plasmid vector is incubated with the calcium-treated cells and then plated out onto selective medium. Often the marker function on the plasmid determines resistance to an antibiotic. In this case the cells would be plated onto nutrient agar containing that antibiotic. The conditions of the transformation are such that only a single plasmid molecule enters a bacterial cell. The single transformed cell then grows to give a colony of cells on the plate. Each of these colonies contains a homogeneous population of identical – 'cloned' – plasmids derived from the single transforming plasmid.

The linear DNA of bacteriophage λ vectors also has to be cleaved with a restriction endonuclease before it can accept segments of foreign DNA. The foreign DNA is also usually cleaved with a restriction endonuclease so it can be joined to the vector by the action of DNA ligase. There are two major types of λ vector, which are described in Chapter 3. One type has a single cleavage site into which the foreign DNA is *inserted* (Section 3.1.2(b)). In the other, the vector has two cleavage sites for the restriction enzyme and DNA is inserted between these two sites to *replace* a segment of DNA in the vector (Section 3.1.2(c)). The heterogeneous population of molecules from this reaction mixture can then be introduced into bacteria by the transfection procedure outlined above. There is, however, an alternative strategy whereby λ DNA can be packaged *in vitro* into phage particles which can be used to infect the host cell (Section 3.1.3(c)). The insertion of foreign DNA does not affect the ability of the recombinant molecule to replicate in the host bacterium and to produce progeny recombinant phage particles. The infection of a single bacterial cell by a single recombinant molecule therefore results in the clonal propagation of the recombinant molecule. In this way homogeneous populations of molecules can be prepared from phage preparations made initially by picking a single phage plaque from a bacterial lawn.

The next problem to be faced is how to identify which colony or which plaque contains the DNA sequence one wishes to isolate. The most commonly used screening approach requires that one has a pure or partially pure nucleic acid sequence which can be radiolabelled and used as a 'probe' to identify the corresponding recombinant DNA. The probe may be a partially pure mRNA, a chemically synthesized oligonucleotide or a related gene. The screening method of choice is the 'colony hybridization' technique developed by Grunstein and Hogness [10] suitable for use with plasmid vectors, or the analogous 'plaque hybridization' of Benton and Davis [11] for use with phage. In order to carry out colony hybridization the cells are first plated onto selectives plates. A replica of the colonies is made onto a nitrocellulose filter disc which is placed on the surface of a second plate (see Fig. 1.1). The colonies are allowed to grow on the master plate and the nitrocellulose disc. The nitrocellulose filter is then removed and placed onto blotting paper wet with 0.5 N NaOH solution. The alkali diffuses into the nitrocellulose, lyses the bacteria *in situ* and denatures their DNA. The filter is then neutralized using Tris buffer in the presence of high salt. The single stranded DNA binds to the nitrocellulose filter in the position originally occupied by the bacterial colony. The filter is then baked at 80° C, following which it is incubated with a solution containing the radiolabelled probe under conditions which favour nucleic acid hybridization. Unhybridized material is removed by extensive washing, thus allowing the identification of colonies containing sequences complementary to the probe by autoradiography. Colonies which give a positive autoradiographic signal can then be picked from the master plate and cultured in order to provide sufficient cells from which to make plasmid DNA. A virtually identical procedure is followed for making a replica of plaques generated in a bacterial lawn by bacteriophage λ recombinants. The filter onto which the plaques are transferred is, however, taken immediately to the alkali denaturation step.

In order to prepare plasmid DNA, cells are cultured in nutrient broth in flasks that are shaken vigorously at 37° C. The cells are pelleted by centrifugation and then treated with lysozyme in a buffered isotonic solution. A variety of detergent treatments can then be used to complete the breakage of the cells. These are carried out under conditions which cause bacterial chromosomal DNA and other cell debris to precipitate and plasmid DNA to be left in solution. Significant purification can therefore be achieved by medium speed centrifugation to pellet the chromosomal DNA and other debris.

Finally ethidium bromide is added to this supernatant together with caesium chloride to bring the solution to a density at which the DNA–dye complex will band. The DNA is then banded to equilibrium by ultracentrifugation. Ethidium bromide is an intercalating dye that will bind to both linear and circular DNA. There is, however, a topological restraint upon the amount of dye that can bind to covalently closed circular DNA. The contaminating chromosomal DNA is inevitably fragmented and this linear DNA, together with any nicked plasmid molecules, binds more ethidium and therefore bands at a lower density than the covalently closed circular DNA. Contaminating protein will float to the surface and RNA will pellet during the banding process. The band of supercoiled plasmid DNA can then be collected from the tube and the dye removed by extracting with an alcohol (see Fig. 1.2).

In order to prepare phage, a single plaque has to be picked and taken through a couple of steps of amplification. The first involves the lysis of a small culture of bacteria either in broth or on a single petri dish. The progeny phage from this step are titred and then used to infect a larger culture of cells in the early 'log phase' of their growth. The multiplicity of infection must be such that all the cells in the culture become infected and therefore all undergo lysis. Phage can be concentrated from the nutrient broth containing the lysed cells by precipitation with polyethylene glycol. This precipitate of phage and other cell debris is resuspended in caesium chloride solution and the phage particles purified by banding to equilibrium in the ultracentrifuge. A concentrated suspension of highly purified phage particles can be collected from such a density equilibrium gradient. The phage DNA is then extracted from the phage particles by deproteinization using phenol (see Fig. 1.3).

1.3 Strategies for gene cloning

1.3.1 Cloning by complementation

This approach has been most successful when the DNA to be cloned is from the same species as the host for the recombinant molecules. The approach may be considered as a direct extension of the techniques used in classical microbial genetics. In its simplest form, a 'library' of cloned segments of *E. coli* DNA (a heterogeneous population of recombinant DNA molecules representative of all the sequences in the *E. coli* genome) is introduced into the cells of an auxotrophic strain of *E. coli*. The recipient cells are plated on minimal medium lacking the substance required by the strain and those cells which have acquired a gene that complements the auxotrophic mutation will grow to produce colonies of transformed cells. The power of this

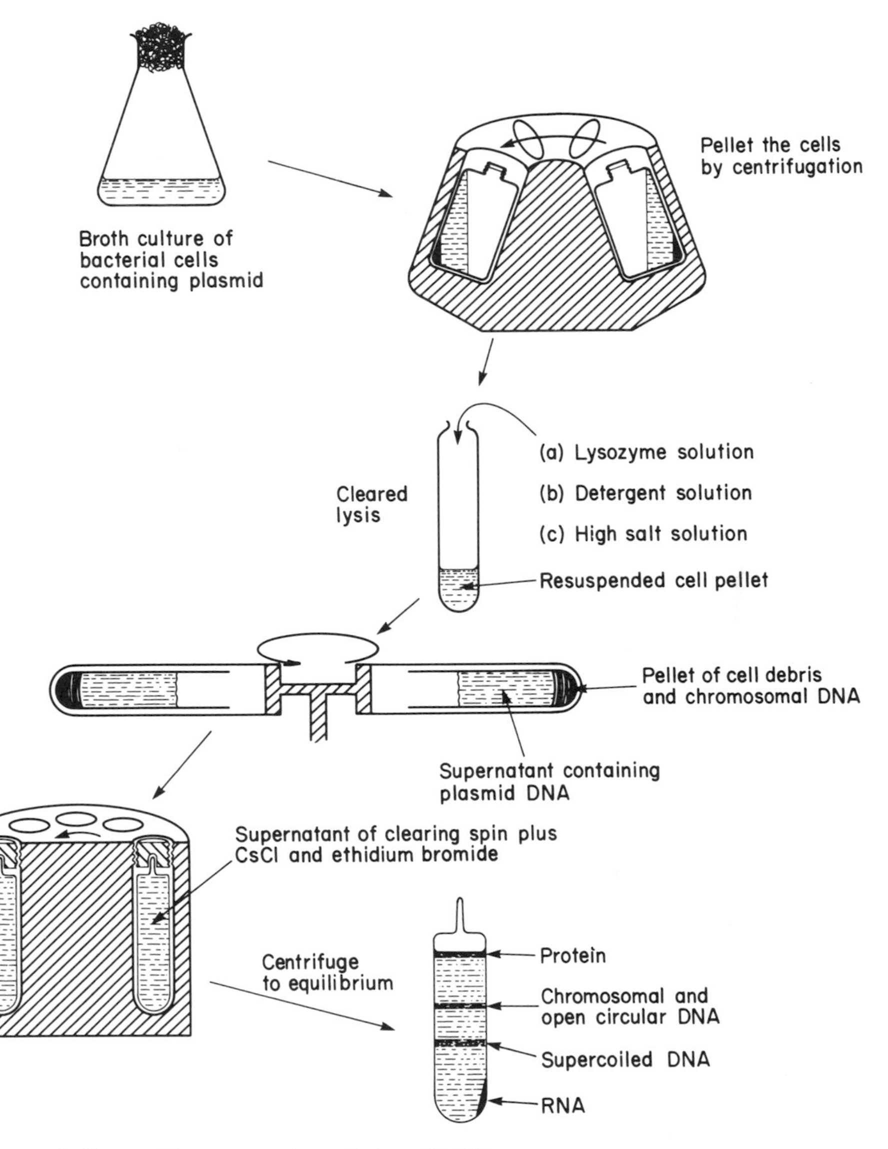

Figure 1.2 The preparation of plasmid DNA.

general approach was impressively demonstrated by Clarke and Carbon [12], who made libraries of random segments of *E. coli* DNA cloned into ColE1 by the dA:dT tailing technique (Section 2.1.3) and successfully isolated portions of several bacterial operons. This direct method of selection requires an auxotrophic strain that can be made competent for transformation. Some strains are, however, difficult to transform and so they developed an alternative approach whereby a

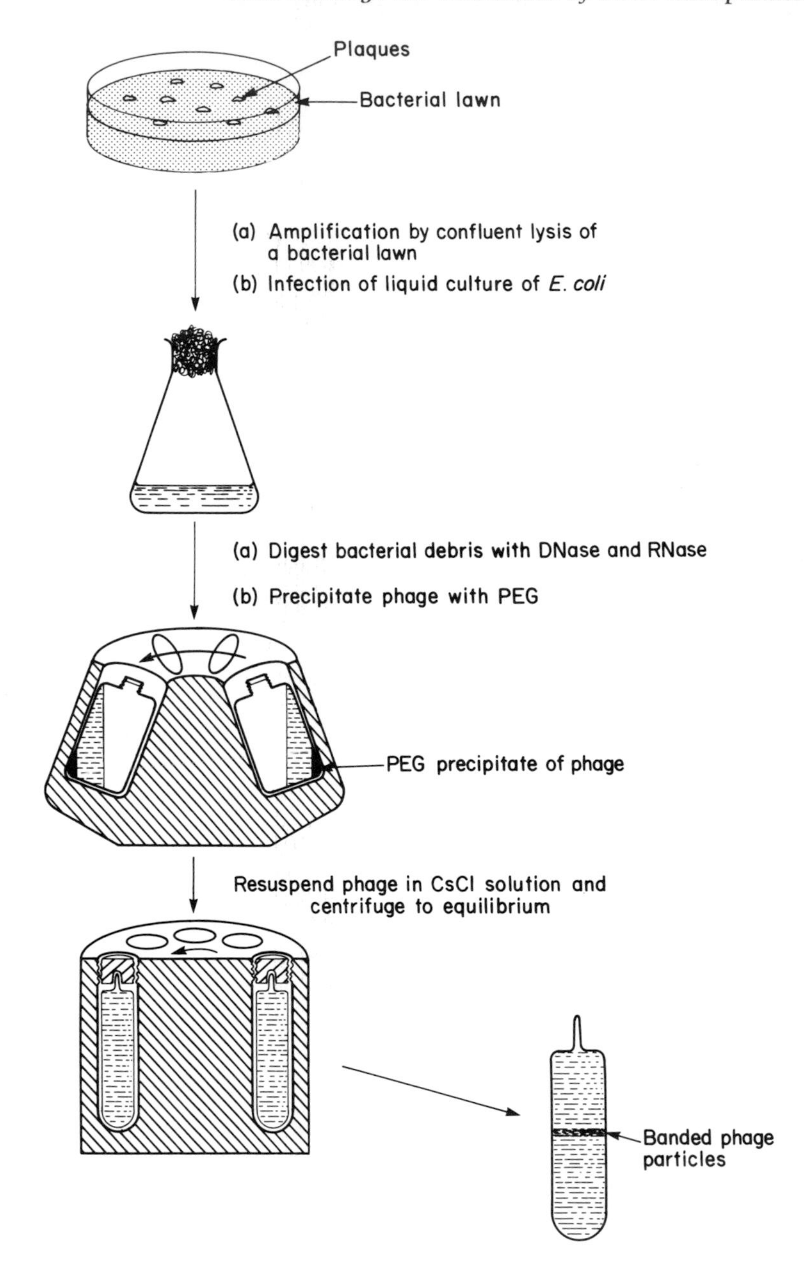

Figure 1.3 The preparation of bacteriophage λ.

pool of recombinant plasmids was first used to transform a strain of *E. coli* harbouring an F plasmid [13]. The recombinant plasmids in their transformants could then be introduced into a number of F^- auxotrophic strains by F-mediated transfer in a replica-mating experiment. In this way it proved possible to isolate many genes from the one library.

The organizational differences between eukaryotic and prokaryotic genes are such that only the chromosomal genes of bacteria and a few genes from lower eukaryotes, such as yeast, can complement *E. coli* auxotrophs. Complementation by the chromosomal DNA of yeast was first shown by Struhl *et al.* [14], who cloned a 10 kb yeast *Eco*RI fragment into a non-revertible *E. coli* histidine auxotroph lacking the enzyme imidazole glycerol phosphate (IGP) dehydratase. In this experiment the yeast DNA, cloned in λ phage, was integrated into the bacterial chromosome using an integration helper phage. Transcription is probably initiated within the yeast DNA, since the λ promoters are either repressed or deleted. The same segment of yeast DNA was also cloned in the ColEl plasmid by Ratzkin and Carbon [15], who also selected recombinants which would complement deletion mutants in the *leuB* gene (β isopropyl malate dehydrogenase) of *E. coli.* The latter plasmid could complement both the *leu* B mutation of *E. coli* and also the equivalent gene in *S. typhimurium.* Subsequently it has been found that about 30% of yeast genes are functional in *E. coli.* These include examples of genes from the biosynthetic pathways of tryptophan (*trp*1), arginine (*arg*8), and uracil (*ura*3). In this respect yeast is an unusual eukaryote in that relatively few of its genes contain intervening sequences.

Attempts have also been made to complement yeast mutants with the DNA from higher eukaryotes. These experiments were undertaken since yeast is a eukaryotic microorganism for which there is a wealth of characterized mutants available. This work did not meet with success, probably because the intervening sequences within yeast genes have a specific sequence that is required for correct splicing and maturation of mRNA. Henikoff and his colleagues [16] were, however, able to isolate a segment of chromosomal DNA from *Drosophila melanogaster* that would complement a yeast *adenine*8 mutation. This segment of DNA is transcribed in yeast to give a 1 kb mRNA that specifies the enzyme glycinamide ribotide transformylase. This corresponds to the *C*-terminal domain of a much larger peptide encoded by a 4.7 kb mRNA in *D. melanogaster*. The original complementation of the yeast mutation was therefore highly fortuitous.

This general problem might be circumvented by cloning cDNAs into expression vectors. These clones could complement mutations in either *E. coli* or yeast. This has been tested using the cDNA cloning technique of Okayama and Berg (see Section 2.2.2) for cDNAs corresponding to mRNAs from yeast [17].

1.3.2 Cloning cDNA

A DNA copy of mRNA (cDNA) can be synthesized using reverse transcriptase, the key enzyme for the replication of the retroviruses (see Chapter 8). The procedures used for the synthesis of cDNA are discussed in detail in Section 2.2.2. Eukaryotic mRNAs can readily be reverse-transcribed into cDNA since they have 3′ poly-A tails onto which one can anneal an oligo-dT primer for the reaction. The cDNA form of a eukaryotic gene is often used to direct the expression of a eukaryotic protein in *E. coli*. It is necessary to use cDNA copies of a gene rather than chromosomal DNA for this purpose since the *E. coli* cell has no means of removing intervening sequences from the primary transcripts of eukaryotic genes. Mature eukaryotic mRNA has, however, already had the intervening sequences removed by the splicing process. The construction of plasmids for the expression of foreign genes is discussed in Chapter 5. cDNA clones also offer a direct route towards cloning the corresponding chromosomal gene. It is relatively easy to clone the cDNA corresponding to a gene that directs the synthesis of a protein found in a highly differentiated tissue. mRNA synthesized from the insulin gene, for example, is highly abundant in pancreatic tissue just as globin mRNA is highly abundant in erythroid tissue and so on. cDNA cloned from these mRNAs can be used to screen libraries of recombinants representative of the total genomic DNA of the organism and in this way the chromosomal genes can be isolated.

(*a*) *Colony hybridization screens*

The strategy taken in cloning cDNA will reflect the abundance of the mRNA corresponding to the gene of interest within a given tissue. In the above examples, the mRNAs are highly abundant and the isolation of a cDNA clone presents no particular problem. Indeed the mRNA for globin can be purified from an RNA preparation enriched for polyadenylated RNA by oligo-dT cellulose chromatography. If such a poly-A^+ enriched RNA preparation from immature erythrocytes is sedimented on a sucrose gradient then the globin mRNA can be recognized as a distinct 9S peak. RNA purified by either sucrose gradient sedimentation or by gel electrophoresis can be used as the starting material for cDNA cloning thereby reducing the number of clones that have to be screened to obtain the correct recombinant. The cloning procedure is in itself the ultimate in purification procedures and so in the interests of avoiding potential degradation of RNA during a large number of manipulations it is often preferred to synthesize cDNA from total poly-A^+ enriched RNA. This first of all generates a heterogeneous population of recombinant molecules containing cDNA. A homogeneous isolate of any one of these recombinants is obtained following the transformation of *E. coli* under conditions in which a single bacterial cell receives a single recombinant molecule. The desired clone has to be selected from the population of transformed bacterial colonies by a screening procedure. One way to do this would be to carry out the colony hybridization procedure of Grunstein and Hogness [10] that was described above (Section 1.2) using gradient purified RNA as the radiolabelled probe. More simply poly-A^+ RNA could be prepared from two different tissues: one in which the differentiated gene product was synthesized and one in which it was not. These RNAs could then themselves be radiolabelled, or radiolabelled single stranded cDNA could be synthesized from them, and used as probes upon duplicate filters taken from one plate of colonies. Colonies which hybridized only to the probe from the differentiated tissue could then be picked from the master plate and subsequently characterized. Definitive identification of the correct clone can be obtained either by sequencing, if the amino-acid sequence of the particular protein is known, or by one of the translation tests described below. One rapid DNA sequencing technique is described in Section 3.2.3. This type of colony hybridization screen eliminates the so called 'house-keeping' genes that are expressed in all tissues irrespective of their state of differentiation. The approach works well for abundantly expressed mRNAs but it is not so practical in looking for rare transcripts. Such experiments are greatly assisted if there is some knowledge of the amino-acid sequences of the protein encoded by the mRNA. It is then possible to chemically synthesize oligonucleotide probes (see Section 2.2.1) for the colony hybridization procedure. The choice of the sequence to be synthesized is largely dictated by the degeneracy of the genetic code. The oligonucleotide sequence is chosen from a region in which the number of degenerate codons is minimal. It is still usually necessary to synthesize several such probes 13–15 nucleotides in length and which correspond to all possible nucleotide sequences that could encode all or part of the chosen sequence of 4–5 amino-acids.

Alternatively synthetic oligonucleotides can be used to prime the synthesis of cDNA. In this role they would take the place of oligo-dT in the reverse transcriptase reaction (see Fig. 2.10). This approach of directing the synthesis of specific cDNA molecules is sometimes used when a cDNA clone is found not to be a full length copy of the mRNA. A synthetic oligonucleotide corresponding to the sequences at the 5′

end of the first clone can be used to prime the synthesis of cDNA from the missing 5′ segment of the mRNA.

(b) Screening by translation

Although *in vitro* translation has been used as a primary screen to identify cloned DNAs, it is now mainly used as a technique to confirm identification of clones. Perhaps the most striking example of a primary translation screen was the isolation of the interferon gene from total leukocyte poly-A^+ RNA [18] (see also Chapter 5). This screen uses oocytes from the toad *Xenopus laevis* in order to translate mRNA. When poly-A^+ RNA from leukocytes is microinjected into the oocytes it is translated. Interferon, which is amongst the proteins that are synthesized, is secreted into the culture medium and can be detected by its anti-viral activity. The interferon message was thought to represent between 10^{-3} and 10^{-4} of the mRNA, and so on the order of 10^4 transformants had to be screened. Rather than screen colonies individually, DNA was prepared initially from 12 pools of 512 transformed colonies. Of these, four pools gave positive signals and were sub-divided for retesting. This procedure was repeated until a single clone was identified. The test involves immobilizing DNA on a solid support and using this to isolate complementary mRNAs by preparative nucleic acid hybridization. The solid support used to bind the DNA can be either nitrocellulose or diazobenzyloxymethyl cellulose paper (see also Section 6.2.2). The mRNA can subsequently be eluted from the DNA–RNA hybrid and microinjected into oocytes. It is also possible to translate mRNA prepared in this way in a cell-free system derived from rabbit reticulocytes or from wheatgerm. Often the translation products are analysed by polyacrylamide gel electrophoresis and definitively identified by immunoprecipitation techniques.

There is a second translation test used to identify cloned DNAs known as ‘hybrid arrested translation’. In this approach the unfractionated mRNA is hybridized with cloned DNA under conditions which favour the production of DNA–RNA duplexes. The mixture is subsequently translated in a cell-free system. The mRNA in the DNA–RNA duplex will not direct the biosynthesis of protein. One therefore looks for the arrest of synthesis of one of the translation products by polyacrylamide gel electrophoresis.

(c) Direct immunological screening techniques

There are a final set of screening methods that are directly analogous to the colony hybridization techniques. Rather than using radiolabelled nucleic acids as probes, however, antibodies are used to identify colonies or plaques which are synthesizing an antigen encoded by the foreign DNA. Sets of vectors have now been developed specifically for use with this screening method. These 'expression' vectors are designed so that the foreign gene they carry is transcribed and subsequently translated within the bacterial cell. These vectors are described in Section 4.1.6.

The first attempts at immunoassays detected visible immunoprecipitates around the plaques or colonies when the specific antiserum was contained in the growth medium. The sensitivity and rapidity of the screen has been greatly increased by the use of radiolabelled antibodies and a solid phase support for the antibody. Bacteria containing recombinant DNA are replica plated onto nutrient agar and, following growth, the cells within the colonies can be lysed by exposure to chloroform vapour. In some cases lysis is brought about by a temperature shift which induces a prophage encoding a thermosensitive repressor. The antibody, linked to solid phase support, is brought into gentle contact with the lysed cells to allow adsorption of antigen to the antibody. This complex can then be detected by incubating the sheets with a second antibody that is radiolabelled (Fig. 1.4). Unreacted antibody can be washed away, and the position of the complex can be determined by autoradiography. This localizes the position of the bacterial cells which

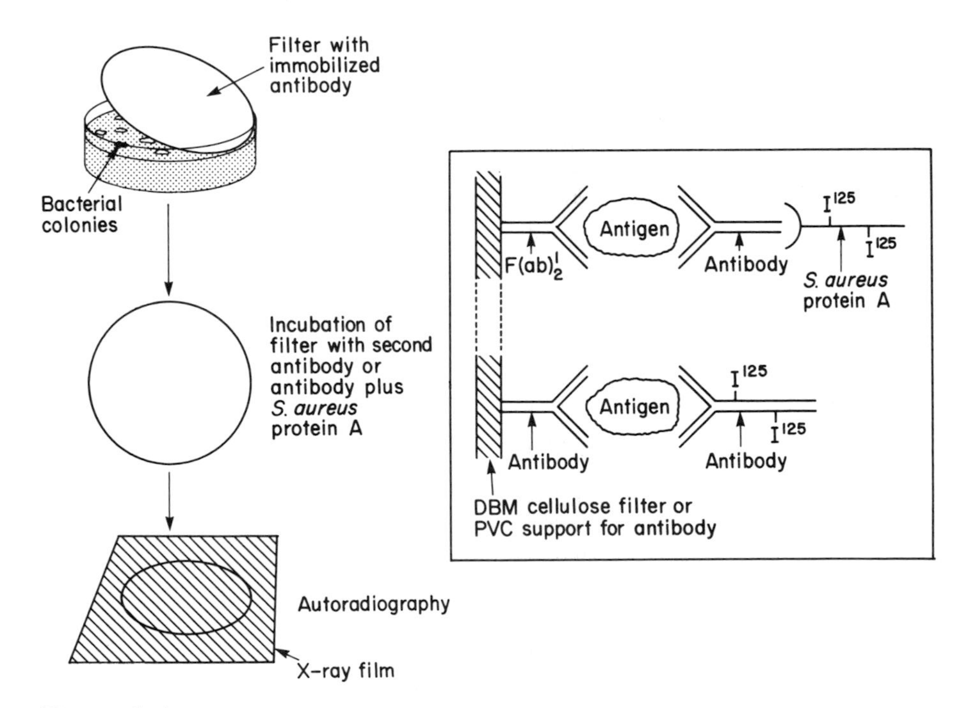

Figure 1.4 Immunoscreening techniques.

are synthesizing the antigen on the master petri dish. In the technique of Broome and Gilbert [19] the antibody is bound to polyvinyl sheets which provide the solid phase support. The same immunoglobulin fraction, but radiolabelled with ^{125}I, is used as a probe for the antigen bound to the immobilized antibody. Ehrlich *et al.* [20] used $F(ab)'_2$ fragments, derived by pepsin digestion of the immunoglobulin, bound to either polyvinyl sheets or diazobenzyloxymethyl cellulose paper. This is incubated first with the antigen and then with undigested antiserum, the FC portion of which will bind radiolabelled *Staphylococcus aureus* A protein (Fig. 1.4).

1.3.3 Cloning genomic DNA

The first step in isolating a complete gene and its flanking chromosomal sequences is often to construct cDNA clones using the approaches described above. The cDNA clone can then be radiolabelled and used as a probe to isolate the chromosomal sequences from a genomic library which should contain cloned sequences representative of the whole genome. Libraries of chromosomal DNA from the higher eukaryotes are usually constructed in bacteriophage λ vectors, although sometimes cosmids are used since they can accommodate larger segments of DNA. The use of λ and cosmid vectors for the construction of libraries is discussed in Section 3.1.4.

It is frequently necessary to analyse long contiguous stretches of chromosomal DNA. This is true for many mammalian genes which are much longer than their corresponding cDNA clones because of multiple intervening sequences. It is not uncommon for single mammalian genes to be isolated as overlapping cloned segments. Related genes are often clustered together in a long chromosomal segment. This is the case for the cluster of the α- or β-globin genes, for example, but the most dramatic cluster so far described contains the genes that determine the histocompatability antigens which in the mouse are present in a 4000 kb segment of DNA that contains on the order of 40 genes. Complex loci are also found in the lower multicellular eukaryotes. The genome of *Drosophila melanogaster*, for example, contains several complex loci, the best studied of which are associated with homeotic mutations that act to control the major pathways which determine development. The bithorax complex, for example, controls the segmental development of thoracic and abdominal structures of the organism. The most dramatic of these mutations result in flies with four wings instead of two, or eight legs instead of six. Most of this gene complex has been cloned and it is found to spread over more than 300 kb. *Drosophila* is extremely well characterized genetically and this facilitates gene isolation. DNA from the bithorax complex was originally isolated by a combination of chromosome 'walking' and chromosome 'jumping'. Chromosome 'walking' involves isolating a series of overlapping cloned chromosomal segments starting out from a particular cloned sequence. Sequences from one end of the starting clone are used to probe a genomic library. The new set of clones isolated from this library are themselves mapped and sequences that are located furthest away from the starting point are then used as probes for yet another screening of the library. If the library is constructed in a λ vector each clone should contain about 15 kb of chromosomal DNA and so one can 'walk' the chromosome in steps of about 15 kb. The progress and direction of 'walks' about the genome of *Drosophila* can be monitored by the procedure of *in situ* hybridization which can localize a nucleic acid sequence to a chromosomal region (see Section 6.3.1). 'Jumping' can be used to bring the destination of the walk closer. It is readily applicable to the *Drosophila* genome where many chromosome translocations and other rearrangements have been mapped. The idea is to choose a chromosome in which the starting point for a chromosomal walk has been brought close to the destination by a chromosomal rearrangement which allows one to 'jump' from one chromosomal region to another. A library of cloned DNAs is then made from flies carrying such a chromosomal rearrangement and screened for overlapping clones. In their isolation of the bithorax complex, Bender, Spierer and Hogness [21] made use of a chromosomal inversion to bring their original clone several thousands of kilobases nearer to the bithorax complex of chromosomal region 89E.

Another technique particularly well suited for cloning genes from *Drosophila* is the approach known as 'transposon tagging'. This was first applied to the isolation of the *white* locus, the X-linked gene concerned with the synthesis of eye pigment. It was known from *in situ* hybridization studies that the mutation *white apricot* (w^a) was associated with the insertion of a transposable element *copia* at the region of the *white* locus. Thus if a library of DNA was constructed from a strain of flies carrying the w^a mutation and this library screened for DNA segments carrying *copia* elements then some of these clones should also contain the *white* locus. The successive screening of such clones by *in situ* hybridization revealed one which carried sequences flanking the *copia* element at the *white* locus [22]. This general principle has been extended in *Drosophila* to mutations caused by hybrid dysgenesis. This phenomenon, which is discussed in detail in Section 8.4.1, results in the insertion of a class of transposable elements, the P-elements, into the mutated genes. A gene marked in this way can be cloned by a procedure analogous to that described above for the cloning of *white*.

1.4 Functional studies with cloned genes

Most studies carried out to date with cloned DNA segments have exploited the cloning technology as a means of preparing large quantities of specific DNA sequences from complex genomes. Detailed physical maps of these specific genes can then be constructed using the techniques described in Chapter 6. Such studies have given us considerable understanding of eukaryotic genomes. One avenue of research which remains to be fully exploited is the use of cloned DNAs to test the functional relationships of DNA sequences, and so to investigate the mechanisms controlling the expression of eukaryotic genes. The general pathway for these investigations is clear. The development of techniques for the *in vitro* recombination of DNA has proceeded alongside the development of techniques for the directed mutagenesis of cloned DNAs. These techniques are examined in Sections 2.3 and 3.2.4. There are numerous possibilities for performing *in vitro* mutagenesis on a cloned sequence with the intention of reintroducing the mutagenized sequence back into the cells of its origin to study the effect of the mutation. It is convenient to carry out most of the cloning and *in vitro* mutagenic techniques using *E. coli* as the host. Now, however, a myriad of vectors have been developed, for introducing cloned DNAs back into the cells of fungi, plants and animals and many of these enable the cloned DNA to be shuttled between the cells of these organisms and those of *E. coli*. These vectors are described in Chapters 7 and 8.

A major application of the *in vitro* recombinant DNA technology which has excited the imagination of industrialists is the potential to design microorganisms which could produce polypeptides of industrial or pharmacological importance. Initially techniques for expressing foreign genes were developed in *E. coli* because of our understanding of the mechanisms by which transcription and translation are controlled in this organism. The development of a family of *E. coli* vectors specifically designed to express foreign genes will be examined in Chapter 5. Some of the ventures to produce mammalian gene products in *E. coli* have been highly successful. Human growth hormone, for example, is produced in high yields as a soluble protein which can be easily purified from large scale cultures of *E. coli*. The product is much cheaper and probably of higher purity than that conventionally used for the treatment of dwarfism, which is extracted from the pituitaries of human cadavers. With other gene products, however, difficulties have arisen from protein degradation, insolubility or lack of post-translational modification. This is causing increased attention to be focussed upon some of the mammalian cell cloning systems as means of expressing cloned mammalian genes (see Chapter 8). The recent success at introducing functional cloned genes back into animals could point to a dramatic expression system with none of the above problems. The production of mice which express multiple copies of the growth hormone gene (see Section 8.4.2) is probably a forerunner to the development of animal stock carrying high copy numbers of genes for pharmacologically important molecules.

References

1. Jackson, D. A., Symons, R. M. and Berg, P. (1972) Biochemical method for inserting new genetic information into DNA of Simian virus 40: circular SV40 DNA molecules containing λ phage genes and the galactose operon of *Escherichia coli*. *Proc. Natn Acad. Sci. USA*, **69**, 2904–9.
2. Lobban, P. and Kaiser, A. D. (1973) Enzymatic end to end joining of DNA molecules. *J. Mol. Biol.*, **78**, 453–71.
3. Merz, J. E. and Davis, R. W. (1972) Cleavage of DNA by R1 restriction endonuclease generates cohesive ends. *Proc. Natn Acad. Sci. USA*, **69**, 3370–4.
4. Berg, P., Baltimore, D., Brenner, S., Roblin, R. O. and Singer, M. F. (1975) On the dangers of genetic meddling. *Science*, **192**, 938–40.
5. Revised NIH Guidelines (1979) *United States Federal Register*, **44**, 69210.
6. Williams, R. E. O. (1976) Report of the Working Party on the Practice of Genetic Manipulation. *Command 6600*, Her Majesty's Stationery Office, London.
7. Curtiss, R., Pereira, D. A., Clark, J. E. *et al.* (1976) in *Proceedings of the Tenth Miles Symposium*, Raven Press, New York.
8. Mandell, M. and Higa, A. (1979) Calcium dependent bacteriophage DNA infection. *J. Mol. Biol.*, **53**, 159–62.
9. Hanahan, D. (1983) Studies on transformation of *Escherichia coli* with plasmids. *J. Mol. Biol.*, **166**, 557–80.
10. Grunstein, M. and Hogness, D. S. (1975) Colony hybridisation: a method for the isolation of cloned DNAs that contain a specific gene'. *Proc. Natn Acad. Sci USA*, **72**, 3961–5.
11. Benton, W. D. and Davis, R. W. (1977) Screening λgt recombinant clones by hybridisation to single plaques *in situ*. *Science*, **196**, 180–2.
12. Clarke, L. and Carbon, J. (1975) Biochemical construction and selection of hybrid plasmids containing specific segments of the *Escherichia coli* genome'. *Proc. Natn Acad. Sci USA*, **72**, 4361–5.
13. Clarke, L. and Carbon, J. (1976) A colony bank containing synthetic ColEl hybrid plasmids representative of the entire *E. coli* genome. *Cell*, **9**, 91–9.
14. Struhl, K., Cameron, J. R. and Davis, R. W. (1976) Functional genetic expression of eukaryotic DNA in *Escherichia coli*. *Proc. Natn Acad. Sci USA*, **73**, 1471–5.
15. Ratzkin, B. and Carbon, J. (1977) Functional expression of cloned yeast DNA in *Escherichia coli*. *Proc. Natn Acad. Sci. USA*, **74**, 5041.

16. Henikoff, S., Tatchell, K., Hall, B. D. and Nasmyth, K. A. (1981) Isolation of a gene from *Drosophila* by complementation in yeast. *Nature*, **289**, 33–7.
17. McKnight, G. L. and McConaughy, B. L. (1983) Selection of functional DNAs by complementation in yeast. *Proc. Natn Acad. Sci. USA*, **80**, 4412–6.
18. Nagata, S., Taira, H., Hall, A. *et al.* (1980) Synthesis in *E. coli* of a polypeptide with human leukocyte interferon activity. *Nature*, **284**, 316–20.
19. Broome, S. and Gilbert, W. (1978) Immunological screening method to detect specific translation products. *Proc. Natn Acad. Sci. USA*, **75**, 2746–9.
20. Erlich, H. A., Cohen, S. N. and McDevitt, H. O. (1978) A sensitive radioimmunoassay for detecting products translated from cloned DNA fragments. *Cell*, **13**, 681–9.
21. Bender, W., Spierer, P. and Hogness, D. S. (1983) Chromosomal walking and jumping to isolated DNA from the *Ace* and *rosy* loci and the bithorax complex in *Drosophila melanogaster*. *J. Mol. Biol.*, **168**, 17–34.
22. Bingham, P. M., Lewis, R. and Rubin, G. M. (1981) Cloning of DNA sequences from the white locus of *D. melanogaster* by a novel and general method. *Cell*, **25**, 693–704.

2 Recombination and mutagenesis of DNA *in vitro*

2.1 The enzymology of *in vitro* DNA recombination

2.1.1 Restriction endonucleases

The ease with which DNA molecules can now be joined *in vitro* is a consequence of the availability of restriction endonucleases, enzymes which recognize specific sequences in DNA and then cleave both strands of the duplex. These enzymes have been found in many prokaryotes and are likely to be responsible for the degradation of 'alien' DNA molecules, the indigenous DNA being protected from degradation by a modification enzyme, usually a methylase. Restriction endonucleases are responsible for the phenomenon of host controlled modification of bacteriophage, first described in the early fifties (reviewed in [1]). If phage λ, which has been propagated on *E. coli* strain K, is then allowed to infect *E. coli* strain B, the efficiency of the infection is very low. The progeny phage can, however, reinfect *E. coli* strain B with high efficiency (Fig. 2.1). Three genetic loci can be identified which control this system: *hsd*S, *hsd*M and *hsd*R. A polypeptide which governs the specificity of the system is determined by *hsd*S. The gene product of *hsd*M is the modification enzyme which also interacts with the product of the *hsd*R gene, the restriction endonuclease, in the cleavage process. In the above examples the phage grown on strain K would have been modified at sites specified by the K restriction–modification system. In the first infective cycle in *E. coli* B cells the B restriction–modification system detects the absence of B modification and degrades the infecting DNA. A small proportion of molecules are, however, methylated by the B modification system, and these survive restriction on the next infective cycle. This phenomenon has to be borne in mind when introducing foreign, unmodified, *in vitro* recombinant DNA into *E.*

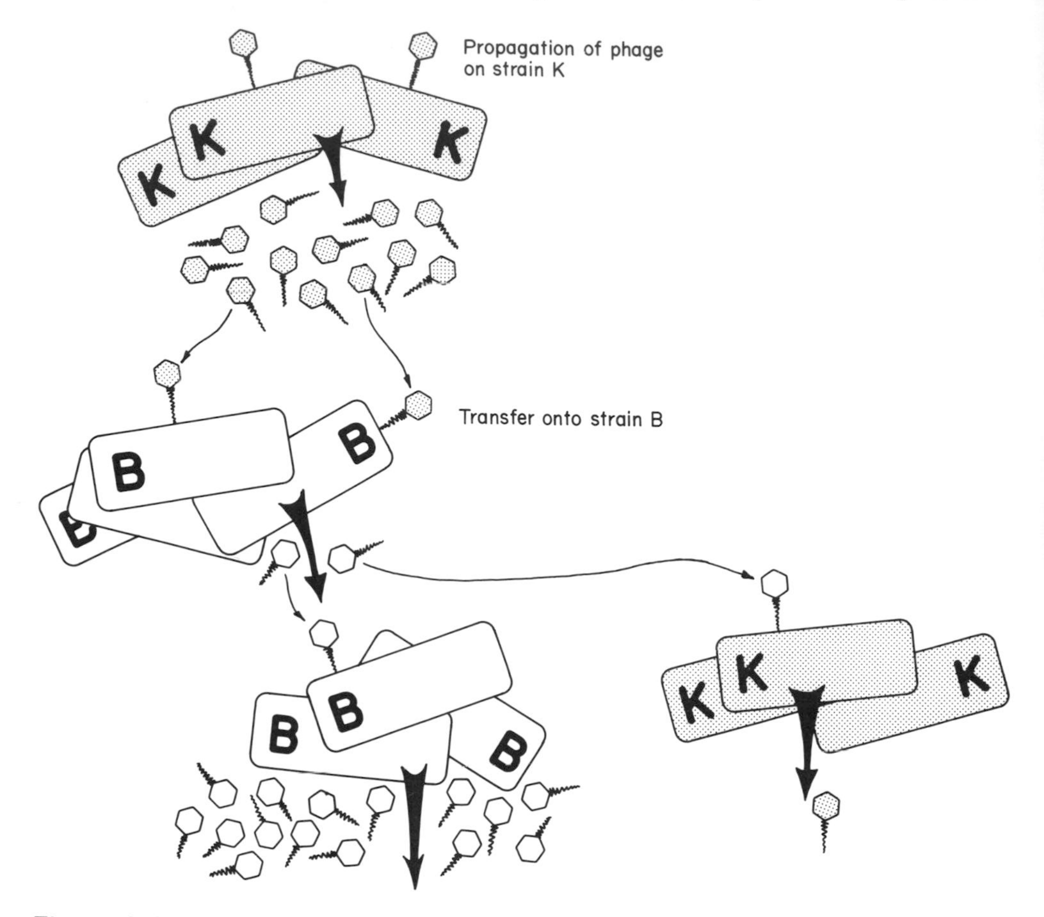

Figure 2.1 The phenomenon of host-controlled restriction and modification.

coli. In order for these molecules to survive, the recipient strain should have defective *hsd*S or *hsd*R genes.

The B–K restriction–modification enzyme systems of *E. coli* have been termed class 1 enzymes: they require Mg^{2+}, S-adenosylmethionine and ATP as cofactors, and although they recognize specific sites within the DNA they do not cleave at these sites [2]. A second class of restriction endonucleases has been identified which have simple cofactor requirements and cleave DNA at, or near, specific sequences that are usually several nucleotides long and rotationally symmetrical about the central nucleotide pairs. These latter enzymes have been isolated from a wide range of prokaryotic microorganisms and are invaluable for cloning DNA.

An extensive list of these enzymes, and the sequences recognized by them can be found in reference [3]. Let us only consider the recognition sites of some of the more commonly used restriction endonucleases. In general, these enzymes cleave DNA to generate a nick with a 5′ phosphoryl and 3′ hydroxyl terminus. In some cases the cleavages in the two strands are staggered and because of the symmetry of the recognition sequence this generates mutually cohesive termini. The plasmid encoded *E. coli* enzyme *Eco*RI is an example of such an enzyme, and cleaves the sequence GAATTC between the G and A residues (see also Fig. 2.2). In the case of *Eco*RI the protruding single stranded ends have 5′ termini. Other enzymes such as *Pst*I, isolated from *Providencia stuartii*, have a staggered cleavage that generates single stranded 3′ termini. There are also enzymes such as *Hae*III from *Haemophilus aegyptius* which generate flush ends. The type II restriction endonucleases provide the means for dissecting simple genomes or cloned segments from complex genomes. These applications of restriction endonucleases will be summarized in Chapter 6. This chapter will concentrate upon the uses of these enzymes for *in vitro* DNA recombination.

2.1.2 Joining restriction fragments with DNA ligase

The most widely used procedure for recombining DNA *in vitro* makes use of those restriction enzymes which generate mutually cohesive termini on DNA. This type of scission was first recognized by Mertz and Davis [4] who showed by electron microscopy that *Eco*RI-cleaved

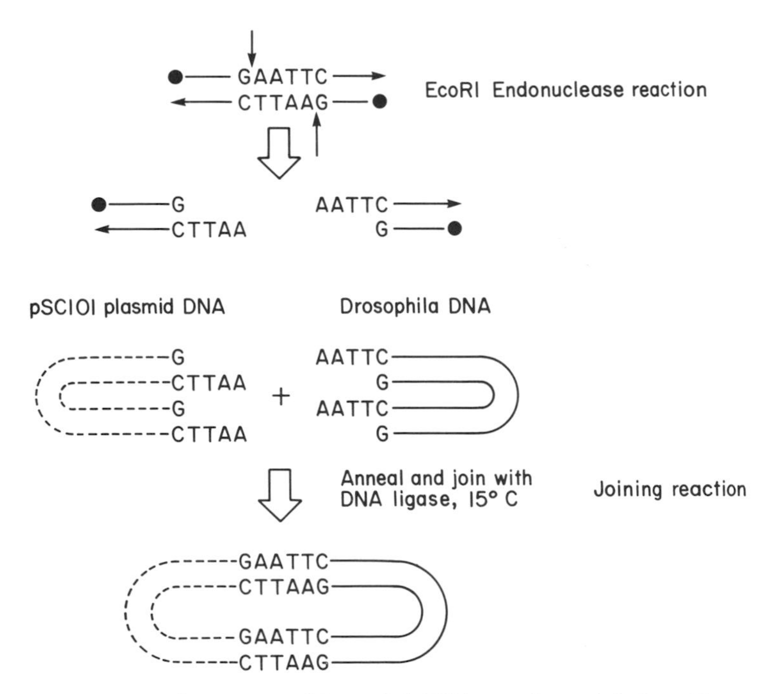

Figure 2.2 Joining *Eco*RI fragments of *Drosophila* DNA to a bacterial plasmid.

DNA would cyclize at low temperature. Furthermore they were able to show that the cohesive termini could be covalently sealed with *E. coli* DNA ligase and were able to construct recombinant DNA molecules between the bacterial plasmid *λdvgal* and the DNA of SV40.

Table 2.1 Commonly used restriction endonucleases which generate cohesive termini

Enzyme	Microorganism	Cleavage site
*Eco*RI	Escherichia coli RY13	G↓AATTC
*Hin*dIII	Haemophilus influenzae Rd	A↓AGCTT
*Sal*I	Streptomyces albus G	G↓TCGAC
*Bam*HI	Bacillus amyloliquefaciens II	G↓GATCC
*Bgl*II	Bacillus globigii	A↓GATCT
*Mbo*I	Moraxella bovis	↓GATC
*Sau*3A	Staphylococcus aureus 3A	↓GATC
*Pst*I	Providencia stuartii 164	CTGCA↓G

A number of enzymes are now known to produce cohesive termini (Table 2.1). Bacteriophage or plasmid vectors have been described which will permit the cloning of fragments generated by many of these enzymes. Some of these enzymes share common central tetranucleotides in their recognition sequence e.g. *Bam*HI, *Bgl*II, *Sau*3A and *Mbo*I, and so although these enzymes recognize different sites in DNA, they will produce identical single stranded 5′ tails which allow the joining of fragments generated by the different enzymes within this set. The identical nature of the termini of DNA fragments from any organism following restriction endonuclease cleavage is the very property which permits the annealing and subsequent ligation of DNA from diverse sources. The general principle of this cloning approach is illustrated in Fig. 2.2 for the specific case of cloning *Eco*RI fragments of *Drosophila* DNA in the bacterial plasmid pSC101.

DNA ligase has the physiological role of sealing single strand nicks in DNA which have 5′ phosphoryl and 3′ hydroxyl termini. Such nicks are generated in repair processes and are also present between the discontinuous DNA segments in replication forks. The two enzymes which are extensively used for covalently joining restriction fragments are the ligase from *E. coli* and the enzyme encoded by phage T4. The *E. coli* enzyme uses NAD as cofactor whereas the T4 enzyme uses ATP, but in either case the cofactor serves to adenylate the ε-NH_2 group of a lysine residue in the enzyme. The 5′ phosphoryl terminus of the DNA is then adenylated by the enzyme cofactor complex and finally a phosphodiester bond is formed with the liberation of AMP (Fig. 2.3) [5]. The enzyme purified from T4 infected

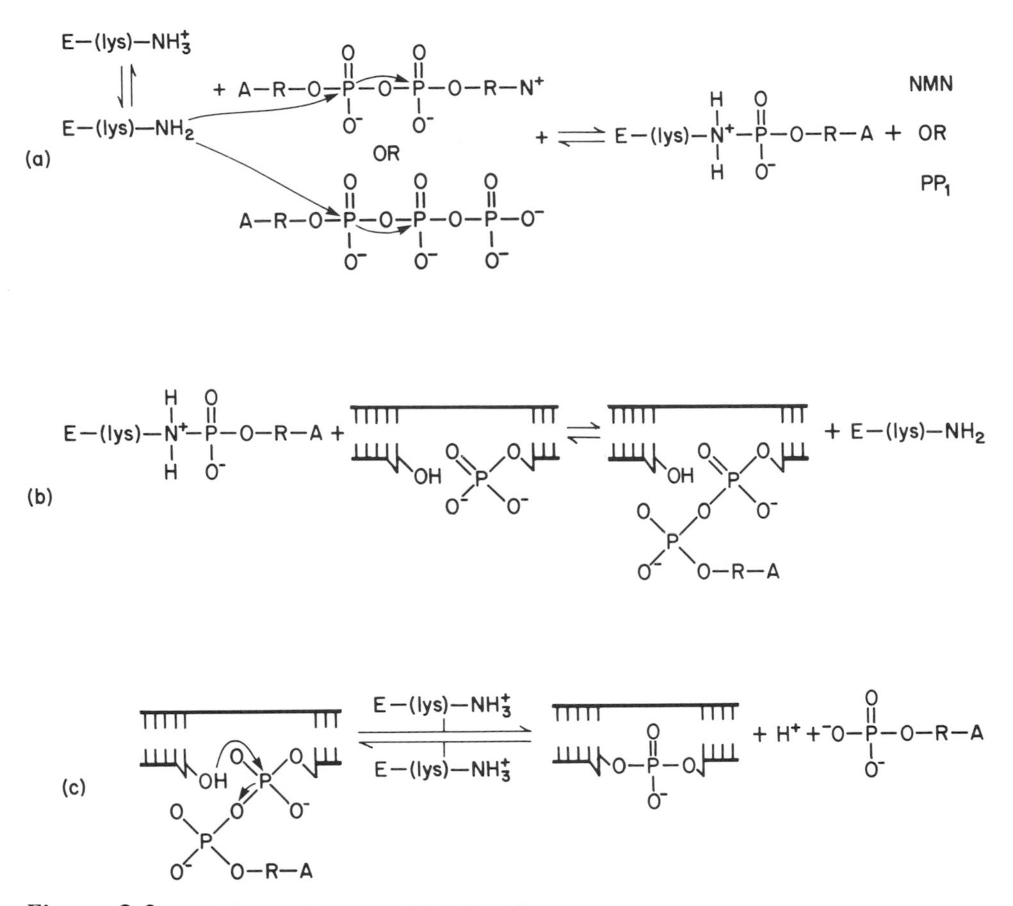

Figure 2.3 The mechanism of the DNA ligase reaction.

E. coli has been used most extensively since it is easier to prepare. The T4 enzyme has the additional advantage that, with high concentrations of enzyme and ATP, it will join DNA molecules cleaved by restriction enzymes which generate fully base paired 'flush ends'. In this case the molecules to be joined are not held together by hydrogen bonds between mutually cohesive termini [6]. The ease by which this enzyme can be prepared has been facilitated by the cloning of its gene in phage λ vectors (see Section 3.1.5).

One major disadvantage of joining a plasmid vector to foreign DNA at cohesive ends generated by restriction endonucleases is the frequency of self-cyclization of the vector plasmid. This results in a 'background' of transformed colonies which contain only the vector plasmid. This can be overcome by treating the restricted plasmid with either bacterial or calf intestinal alkaline phosphatase in order to remove the terminal 5′ phosphoryl groups. The two ends of the plasmid vector can then no longer be covalently joined by DNA ligase. The restriction fragments of the foreign DNA are not, however, treated with phosphatase and so their 5′ phosphoryl groups can be covalently joined to the 3′ hydroxyl groups of the plasmid (Fig. 2.4). This results in hybrid molecules in which, at each site of ligation, the vector is joined to the foreign DNA in one strand only whilst the other strand has a nick with 3′ and 5′ hydroxyl groups. Such a molecule can be introduced into the bacterial cell whereupon these nicks are repaired.

Another major disadvantage of joining DNAs at restriction sites is often encountered when the interest lies in either cloning a large polypeptide coding sequence or a large segment of chromosomal DNA which contains several restriction sites. One early approach which was used to get around this problem was to clone partial *Eco*RI digests of chromosomal DNA in a bacterial plasmid [7]. This is quite a laborious technique because the partial digestion products and the ligated *Eco*RI fragments have to be carefully sized in order to discriminate against clones containing oligomers of restriction

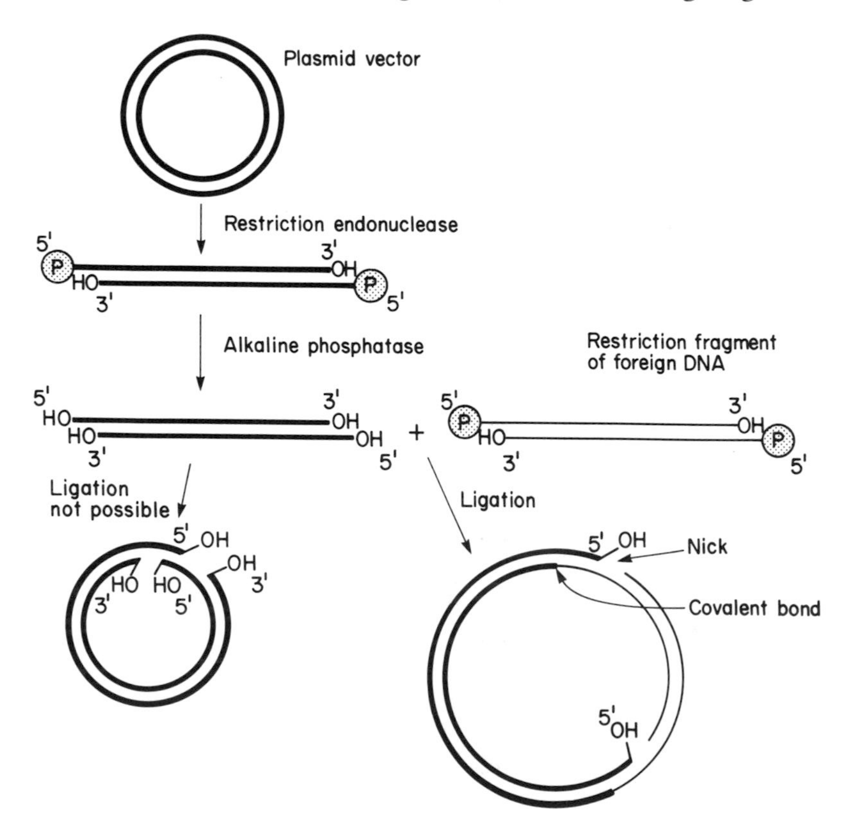

Figure 2.4 Dephosphorylation to prevent the recyclization of linear plasmid DNA.

fragments which were not originally adjacent to each other in the chromosome. There is no way around the application of such careful sizing steps, although some of the other shortcomings of this earlier method have been corrected in a general approach adopted by Maniatis and coworkers [8], who set out to build 'libraries' of cloned DNAs representative of the genomes of several higher organisms (Fig. 2.5). The distribution of restriction sites within the genome of an organism is not random, but is determined by the functional arrangements of nucleotide sequences, and so the cloning of DNA digested by an endonuclease such as *Eco*RI could lead to selective exclusion of DNA from pools of recombinants. This is exacerbated by the size limitations imposed upon recombinants constructed in the λ vectors used to construct such libraries. The packaging of DNA into the capsid places upper and lower size limits upon the length of foreign DNA that can be cloned into the phage vectors (see Section 3.1.2(a)). Very long *Eco*RI fragments may therefore be excluded from a library constructed of DNA partially digested with

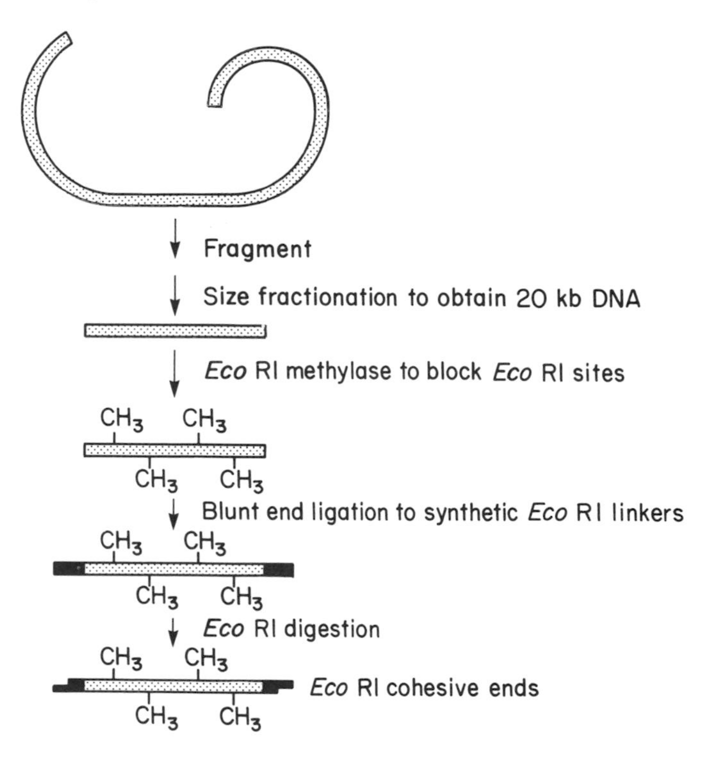

Figure 2.5 The addition of linkers to randomly fragmented chromosomal DNA.

*Eco*RI. In order to overcome this potential problem Maniatis *et al.* [8] generated randomly broken segments of chromosomal DNA by partial digestion with *Hae*III and *Alu*I. These enzymes recognize the tetranucleotide sequences GGCC and AGCT respectively, and cleave DNA to generate flush ends. Specific tetranucleotide sequences occur more frequently in DNA than specific hexanucleotide sequences, and so there is a high probability that any segment of DNA will contain cleavage sites for one of these two enzymes. This partially cleaved DNA is then fractionated by velocity sedimentation on a sucrose gradient and fragments with a mean size of about 20 kb are selected for cloning in a λ phage designed to accept *Eco*RI fragments. This is achieved by adding chemically synthesized oligonucleotides containing the *Eco*RI recognition sequence onto the flush ends generated by *Hae*III and *Alu*I. These oligonucleotides are known as 'linkers'. They consist of self-complementary decameric or dodecameric sequences which self-anneal to give 'flush-ended' DNA fragments which are joined onto the size fractionated chromosomal DNA using T4 ligase. An *Eco*RI cohesive terminus can then be generated by *Eco*RI digestion. In order not to cleave at internal *Eco*RI sites within the chromosomal DNA at this step, the DNA is modified with *Eco*RI methylase before the linkers are added. This procedure is quite laborious as it necessitates a number of enzymatic reactions. The availability of λ vectors that can accept restriction fragments with '*Bam*HI cohesive ends' has greatly simplified the cloning of random segments of chromosomal DNA (see Section 3.2.3). The random cleavage of high molecular weight chromosomal DNA for cloning in these vectors can be achieved by partial digestion with *Sau*3A or *Mbo*I (see Table 2.1). Fragments of about 20 kb are again selected by sucrose gradient sedimentation and can be cloned directly into the vector since they have identical cohesive ends to those generated by *Bam*HI.

2.1.3 Joining DNA via homopolymeric tails

The techniques established by Lobban and Kaiser [9] and Jackson *et al.* [10], were developed at a time when very few restriction endonucleases were available and before it was realized that enzymes such as *Eco*RI generated cohesive ends when they cleaved DNA. These techniques were first successfully used by Wensink *et al.* [11] to produce hybrid plasmids containing *Drosophila melanogaster* DNA that were capable of replicating in *E. coli* (Fig. 2.6). The key steps of the method are still extensively used in cDNA cloning (Section 2.2.2). The method relies upon restriction endonuclease cleavage to open the circular plasmid DNA at a specific, non-essential site. The foreign DNA needs no specific terminal sequence and could be either

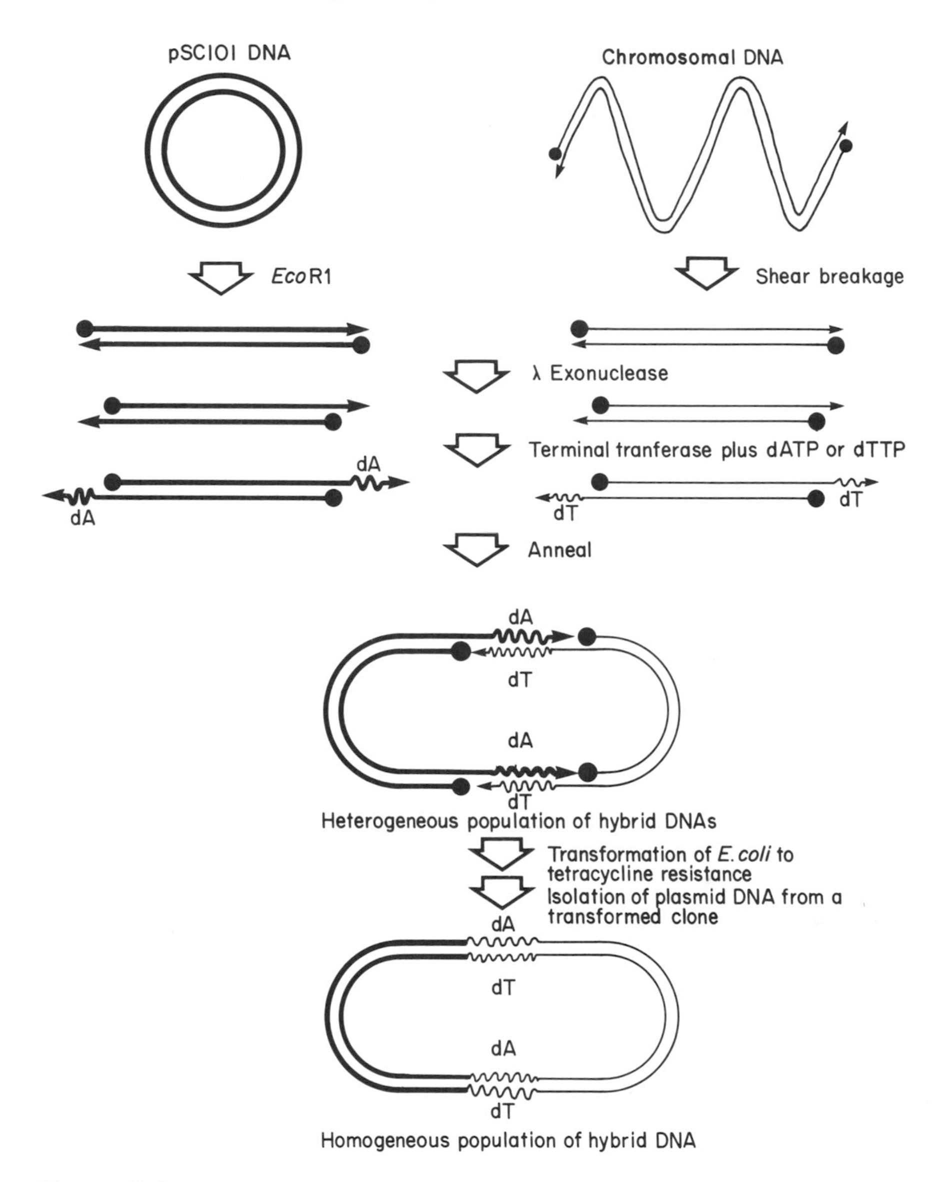

Figure 2.6 The homopolymer tailing technique.

chromosomal DNA that has been fragmented by a number of techniques or a synthetic gene. In the example shown in Fig. 2.6 long chromosomal DNA has been fragmented by hydrodynamic shear to give fragments of 15–20 kb. The linear plasmid molecules and the fragmented foreign DNA are independently treated with λ-exonuclease. This successively removes deoxymononucleotides from the 5′-phosphoryl termini of double-stranded DNA, leaving single-stranded 3′OH termini. Exposed 3′OH termini are good primers for calf thymus terminal transferase, which is used to add homopolymer blocks of deoxyadenylate and deoxythymidylate residues to the respective molecules. The two DNA preparations are then mixed and annealed, whereupon they will join via hydrogen bonding between their homopolymer tails. The original procedure of Lobban and Kaiser [9] then required the covalent sealing of such molecules by the concerted action of exonuclease III, DNA polymerase I and DNA ligase. This procedure was greatly simplified by Wensink *et al.* [11], who introduced the hydrogen bonded molecules directly into *E. coli* and selected transformants which were tetracycline resistant as determined by the plasmid vector pSC101. Roychoudhury *et al.* [12] have described a further simplification of this procedure whereby, using cobalt rather than magnesium as a divalent cation for the terminal transferase reaction, it is possible to add the homopolymeric tails to DNA molecules which have not been treated with λ-exonuclease.

The advantages of this joining method for cloning chromosomal DNA are that one cannot get indiscriminate joining of DNA segments which originate from non-contiguous regions, and that it can be applied to randomly broken DNA segments. Furthermore, vector plasmid DNAs that have homopolymeric single strand extensions cannot self-cyclize, consequently almost all of the bacteria transformed by DNA joined in this manner contain hybrid molecules. One disadvantage of the procedure is that it is difficult to cleave the cloned segment of DNA from the vector DNA in the hybrid plasmid. This is sometimes possible by S1 nuclease digestion, since S1 cleaves single-stranded nucleic acids and will attack breathing duplexes in AT-rich regions. Polydeoxyguanylate and polydeoxycytidylate tails can also be added to DNA molecules following the same procedures. If poly dG tails are added onto a plasmid vector cleaved with *Pst*I, the poly dC-tailed DNA cloned in such a plasmid can often be removed by *Pst*I cleavage, since this joining procedure regenerates the *Pst*I site (Fig. 2.7).

2.2 The synthesis of DNA for cloning

2.2.1 The chemical synthesis of DNA

The use of chemically synthesized DNA in order to 'link' two DNA elements together was described in the previous section. Synthetic decanucleotide and dodecanucleotide linkers containing the recognition sequences of a number of restriction endonucleotides are now generally available. Chemically synthesized oligonucleotides are also used as probes to isolate specific genes (Section 1.3.2(a)). Their

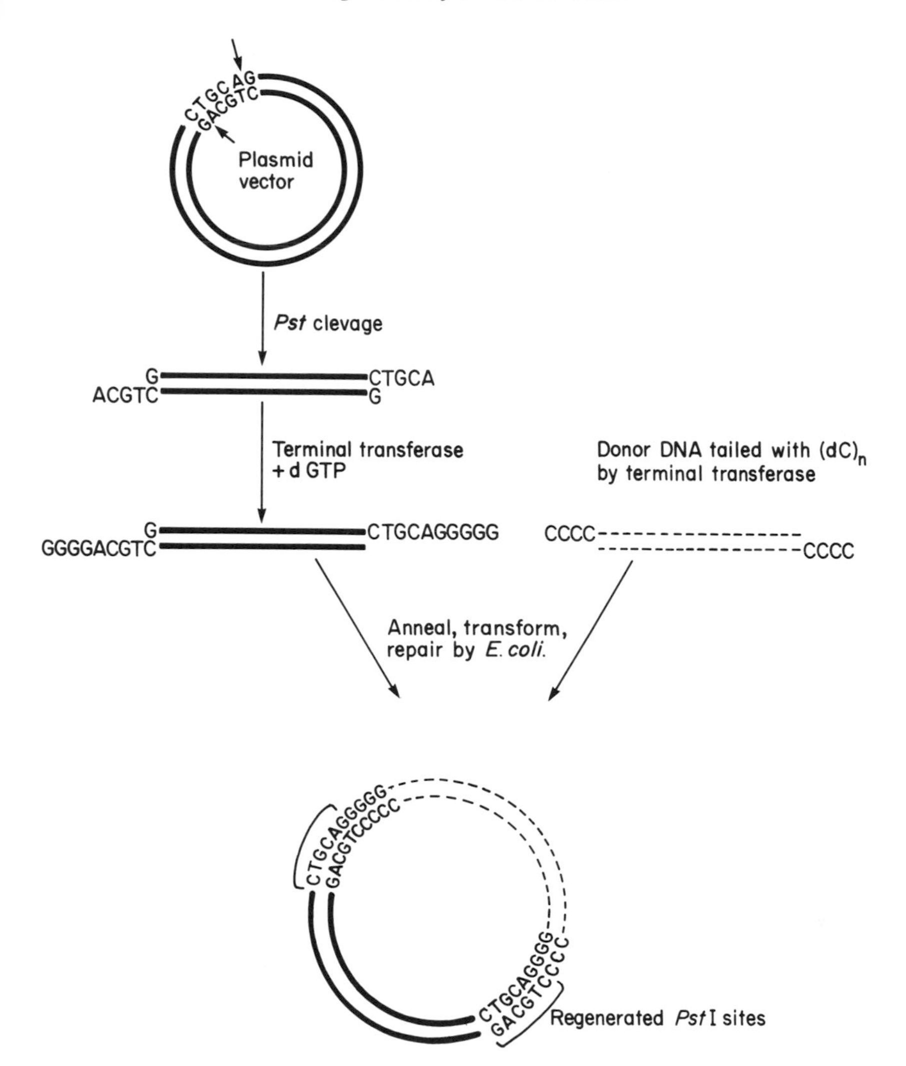

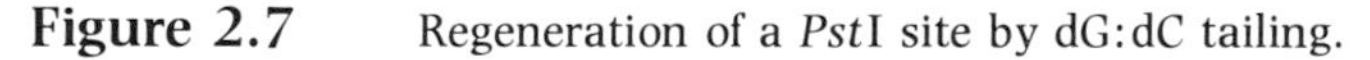
Figure 2.7 Regeneration of a *Pst*I site by dG:dC tailing.

use to direct mutagenesis *in vitro* to particular sequences will be encountered in Section 3.2.4. In fact, entire genes have been constructed from multiple segments of chemically synthesized oligonucleotides. The original 'phosphotriester' method for the chemical synthesis of DNA is complicated and quite laborious as all the reactions are carried out in solution. Nevertheless it was this approach that was used to prepare the gene for somatostatin, the first chemically synthesized gene for a peptide hormone to be expressed in *E. coli* (Section 5.2.1). Subsequently methods have been developed by which the oligonucleotides are synthesized on solid supports such as polystyrene or polyacrylamide [13]. The scheme for one of these methods is shown in Fig. 2.8. The 3′ phosphodiester compound (labelled 'A') is activated by the coupling reagent to react with the 5′OH group of the immobilized nucleoside 'B'. Any unreacted nucleoside has to be capped to give compound 'C'. The main product, 'D', of the coupling reaction is deblocked to give a new 5′OH group on the growing oligonucleotide to which more residues can be added by additional rounds of the reaction. When the synthesis is complete, all the protecting groups are removed and the product is purified by high performance liquid chromatography (HPLC).

An alternative method has been developed in which the initial coupling generates an internucleotide phosphite bond. This is then oxidized to give a phophodiester bond. This reaction is carried out on a silica gel support and is said to be more rapid, with each cycle taking only 30 min. On the other hand a cycle of the phosphotriester method is completed within 60 min and so this is not a serious consideration. One method probably has little advantage over the other.

The original methods for synthesizing genes involved making short oligonucleotides that would hydrogen bond with each other in an unambiguous way. These could then be covalently joined by the action of DNA ligase. In this way it has been possible to synthesize genes as large as a 514 base pair interferon gene, which was assembled from 66 oligonucleotide blocks. An alternative is to synthesize oligonucleotides that anneal to give a structure with large gaps (Fig. 2.9). These gaps can then be repaired with DNA polymerase and DNA ligase.

2.2.2 The synthesis of cDNA

A gene can also be synthesized enzymatically by making a cDNA copy of mature mRNA. This is often the first step in cloning a eukaryotic gene expressed in a differentiated tissue (see Section 1.3.2). It is also the step that is frequently followed to obtain a eukaryotic gene in a form in which it can be expressed in the bacterial cell. The basic technique for cloning cDNA is exemplified by the work of Maniatis *et al.* [14] who were one of the first groups to clone a cDNA copy of the rabbit β-globin gene (Fig. 2.10). Synthesis of the first strand is primed from oligo-dT annealed to the poly-A tail on the 3′ terminus of the mRNA. The RNA template is then hydrolysed and the second strand synthesis primed from hairpin structures which are formed at the 3′ terminus of the first strand. Second strand synthesis can be carried out using either DNA polymerase or reverse transcriptase. The hairpin can then be cleaved with S1 nuclease and the double stranded cDNA can be cloned in a plasmid vector by the tailing technique.

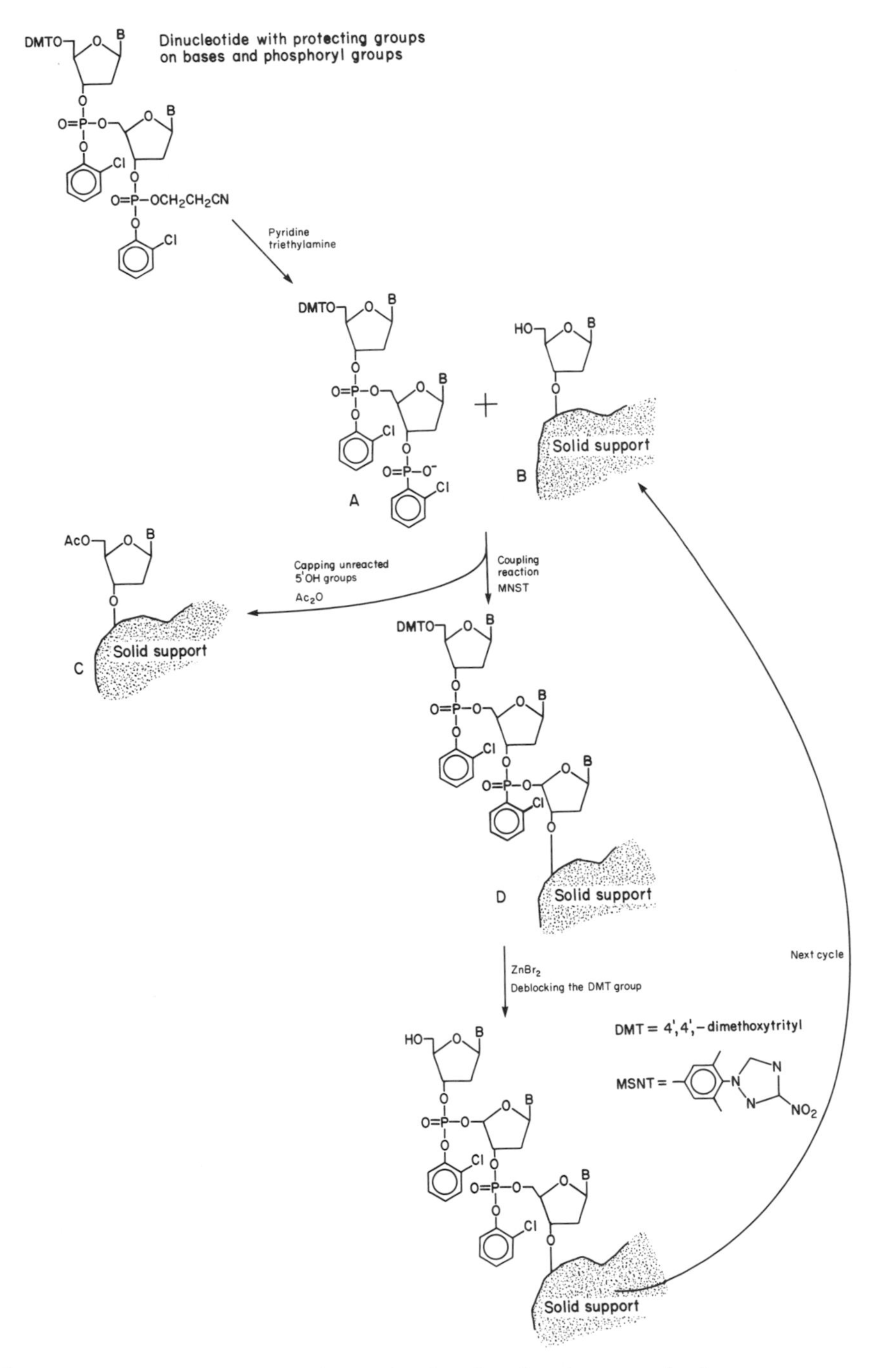

Figure 2.8 Solid phase synthesis of DNA by the phosphotriester method.

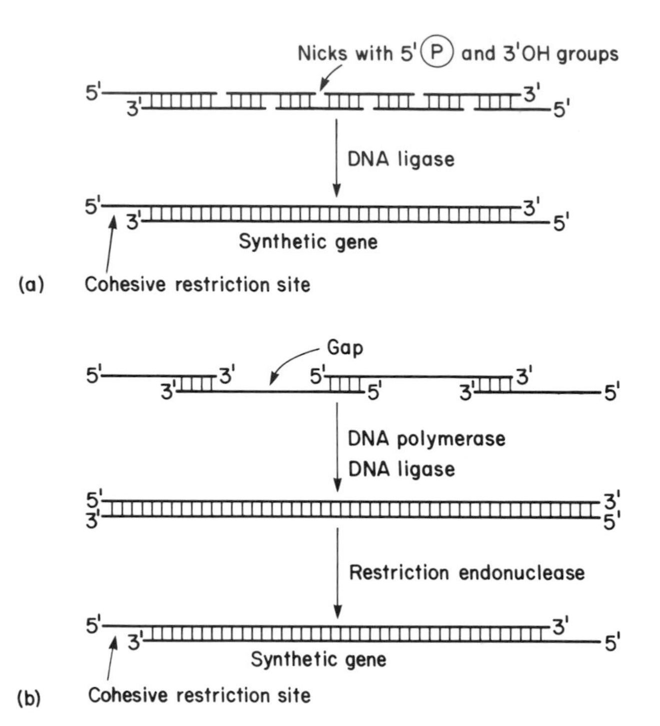

Figure 2.9 The synthesis of genes from blocks of oligonucleotides.

Alternatively synthetic linker sequences containing restriction endonuclease recognition sites may be added to the cDNA using DNA ligase. The linkers are then cleaved with the restriction endonuclease and then the cDNA fragments ligated to the vector. In the example shown in Fig. 2.10, the vector is a bacterial plasmid. Phage vectors are also available for cDNA cloning. Some of the insertion vectors of bacteriophage λ that are described in Section 3.1.2(b) are highly suited for this purpose.

It is technically extremely difficult to generate cDNA clones representative of the full length of a given mRNA. Most cDNA clones contain sequences corresponding to the 3′ end of the mRNA. This is due to several factors including the secondary structure of the RNA itself, which influences the extent of synthesis of the first strand. The second strand reaction is also difficult to control probably because of the highly variable size and position of the terminal loop that acts as a primer for this reaction. Finally the S1 reaction, used to remove this loop, inevitably degrades some terminal nucleotides from the double-stranded cDNA. Okayama and Berg [15] therefore sought to devise a protocol in which full length cDNA could be efficiently cloned. This procedure is shown in Fig. 2.11. Synthesis of the first strand is primed from an oligo-dT tail covalently attached to one end of the linearized

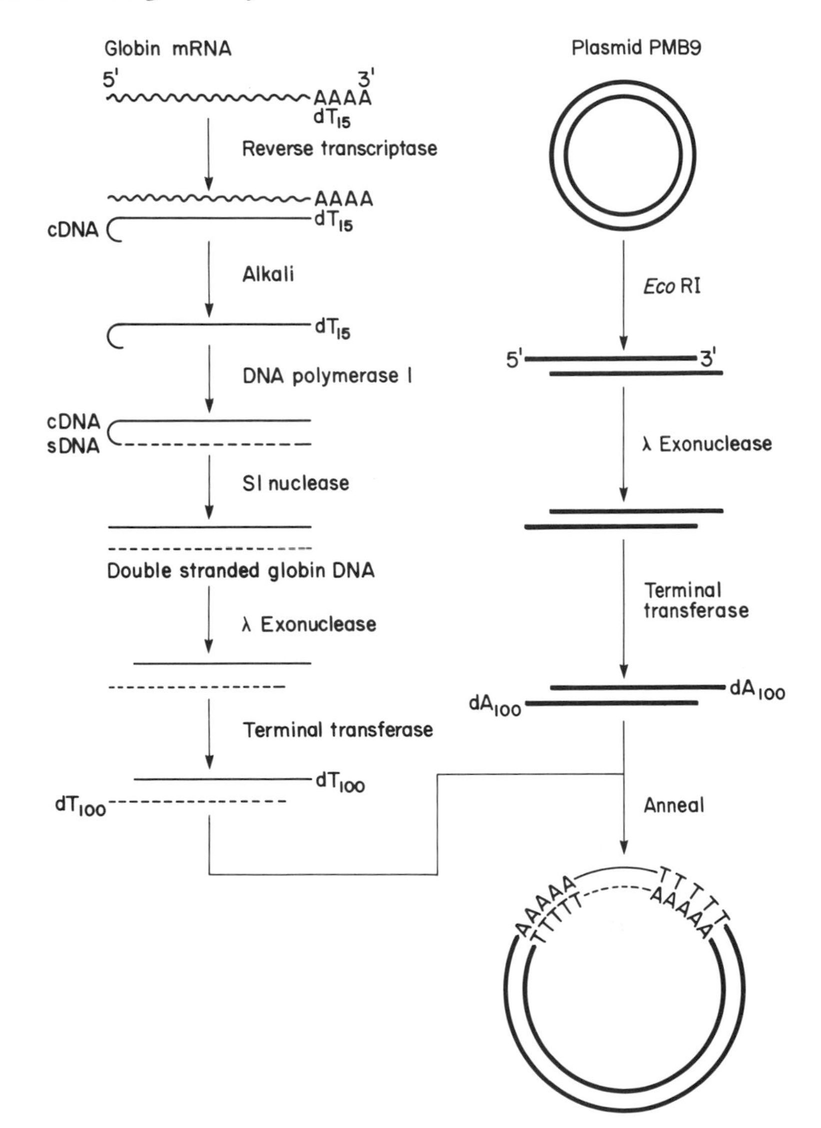

Figure 2.10 Scheme for cloning globin cDNA.

plasmid. The cDNA is therefore immediately attached to the vector in the first step of its synthesis. The synthesis of the second strand is not the next step, instead oligo-dC tails are added to the DNA–RNA duplex. This step probably selects for the full length cDNA product since terminal transferase will preferentially add nucleotides to fully base paired substrates rather than to truncated cDNA transcripts base paired to longer mRNAs. The oligo-dC tailed end of the plasmid vector, opposite to that joined to the cDNA, is removed and replaced

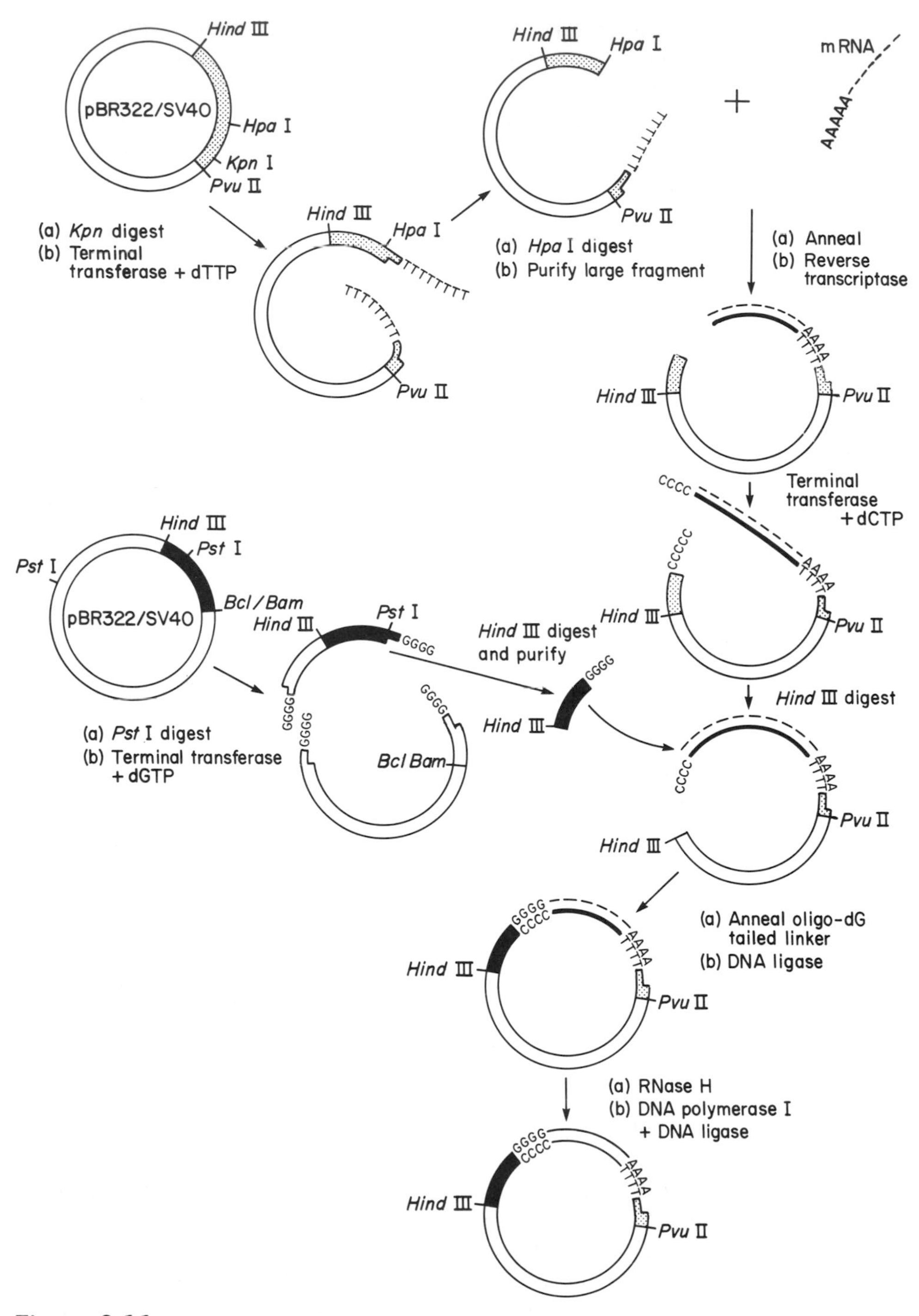

Figure 2.11 The protocol of Okayama and Berg for full length cDNA cloning.

by a similar restriction fragment tailed with oligo-dG. The molecule can then be cyclized. Finally the mRNA is digested from the DNA–RNA duplex using RNaseH. The large gap that is left is repaired using DNA polymerase I and DNA ligase. Several laboratories have now had considerable success in preparing long cDNA clones using this procedure. The protocol promises to be extremely valuable for cloning cDNA adjacent to promoter elements so that the cDNA can be expressed in its recipient cell.

2.3 *In vitro* mutagenesis

2.3.1 Deletion mutants

As a general approach to the study of the function of cloned DNA segments, a number of techniques have been developed to generate mutations at specific sites *in vitro*. The altered DNA segments can then be reintroduced into cells in which the gene is expressed in order to study the effect of the mutation. In several of the chapters which follow, examples will also be given of how restriction fragments within plasmid and phage DNAs have been rearranged during vector development. The simplest way to make deletions is by eliminating restriction fragments. This can be achieved by the consecutive use of the restriction endonuclease and DNA ligase. Examples of this approach can be seen in the development of plasmid vectors described in Chapter 4. In the development of vectors there has been a tendency towards genome diminution that has been achieved in this way. Alternatively, short deletions can be introduced at the cleavage site. Treatment of *Eco*RI generated fragments with S1 endonuclease will remove the single-stranded termini, creating a 'blunt-ended' DNA, so effecting a four base pair deletion at the junction.

More extensive deletions can be made at restriction sites, by treating the cleaved DNA with an exonuclease. This was first demonstrated by Carbon *et al.* [16]. They treated the restricted DNA very briefly with λ-exonuclease to degrade the DNA strands from the 5′ termini and generate protruding single-stranded 3′ termini. These molecules were then introduced into a cell, whereupon they cyclize, presumably as a result of fortuitous homology between the single-stranded termini. Molecules of this kind are evidently repaired *in vivo* by either mammalian or bacterial cells to generate covalently closed circular molecules. These molecules have deletions which extend from the region of fortuitous homology to the restriction endonuclease cleavage site (Fig. 2.12).

The above procedure relies upon the cell to regenerate a covalently closed circular molecule. Cyclization of DNA after exonuclease treatment can also be carried out *in vitro*. A variety of exonucleases

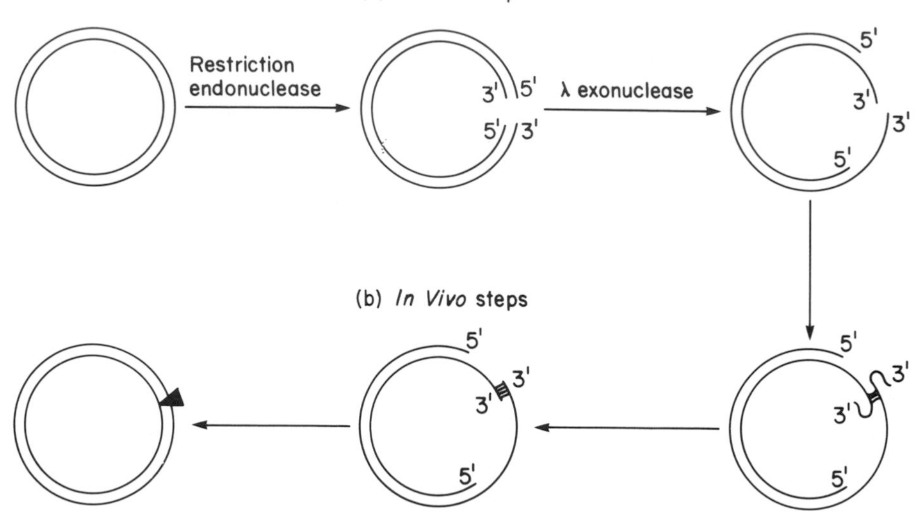

Figure 2.12 Generation of deletion mutants by λ-exonuclease with cyclization *in vivo*.

can be used to generate the deletions. A similar experimental protocol is followed with either λ-exonuclease, exonuclease III, or the exonucleases associated with DNA polymerases. Each of these enzymes leaves the DNA molecule with exposed single-stranded termini which can be degraded by subsequent treatment with S1 nuclease (Fig. 2.13). The resulting 'blunt-ended' molecules can be cyclized using DNA ligase. Exonuclease III is like λ-exonuclease in that it requires double-stranded DNA, but it removes nucleotides in the 3′–5′ direction to leave single-stranded 5′ termini. The 3′–5′ exonuclease associated with the DNA polymerase specified by bacteriophage T4 or the 'Klenow' fragment of *E. coli* DNA

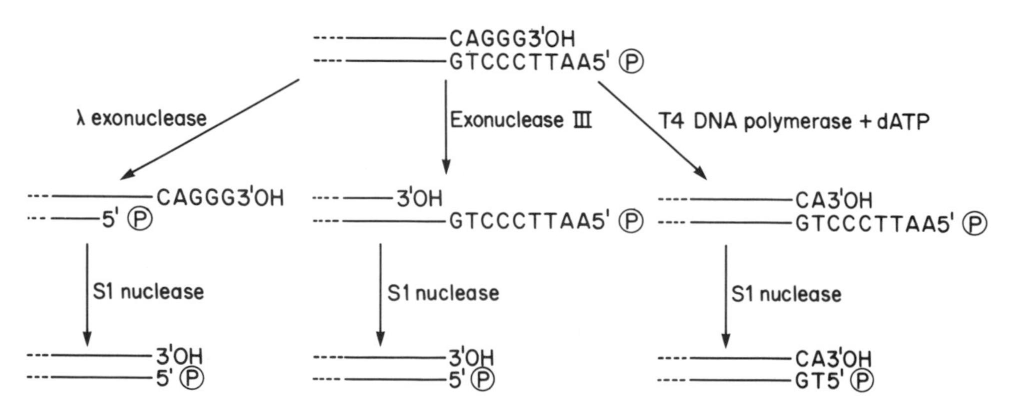

Figure 2.13 Generation of deletions by a combination of exonucleases and S1 nuclease.

polymerase I permit a highly controlled exonucleolytic reaction. If these enzymes are used in the presence of a single deoxynucleoside triphosphate, the 3′–5′ exonuclease activity associated with the polymerase will remove the 3′ terminal residues of double-stranded DNA until the complement of the single deoxynucleoside triphosphate is exposed, whereupon the polymerizing activity will covalently join this residue onto the 3′OH terminus. A stable equilibrium is set up at this point, with a precise number of residues having been removed.

The most commonly used method of generating deletion mutants has been to cut DNA at a unique site and then digest nucleotides away from the ends using *Bal*31 nuclease. This enzyme is both an exonuclease and a single-strand specific endonuclease [17]. It is therefore more convenient to use than either λ-exonuclease or exonuclease III in combination with S1 nuclease, although it is necessary to incubate the *Bal*31 treated DNA with DNA polymerase and the four nucleoside triphosphates so that any remaining 'ragged ends' are filled in. The ends are then usually rejoined via chemically synthesized linkers which create a restriction site at the point of the deletion (Fig. 2.14) [18].

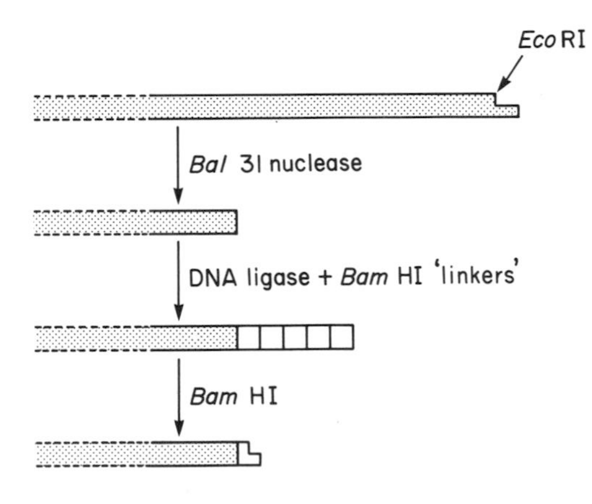

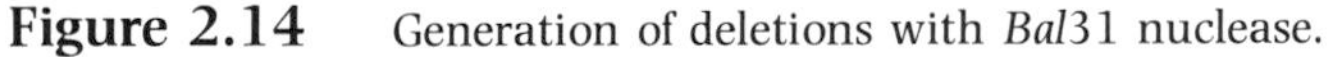

Figure 2.14 Generation of deletions with *Bal*31 nuclease.

Single-stranded DNA can also be exposed to the action of S1 nuclease by catalysing the formation of D-loops with *rec*A protein (Fig. 2.15). D-loops are formed when covalently cloned circular plasmid DNA anneals with a single-stranded DNA fragment complementary to a small region of the molecule. The single-stranded DNA may come from a fragment generated by restriction endonuclease cleavage or it may be a chemically synthesized oligonucleotide. The single-stranded DNA anneals to its complementary sequence in the plasmid, thereby displacing a single-stranded region of the

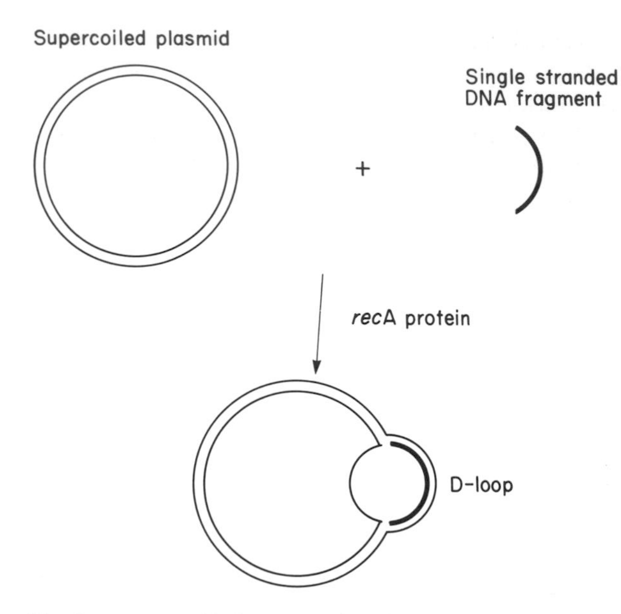

Figure 2.15 The formation of D-loops catalysed by *recA* protein.

plasmid. The D-loop collapses when treated with S1 nuclease which under certain conditions sequentially cleaves both strands of the plasmid DNA. When the resulting linear molecules are religated, they are found to have suffered the deletion of a small number of nucleotides [19].

2.3.2 Insertion mutations

Insertion mutations can be readily generated at the cleavage sites of restriction endonucleases that generate termini with overhanging 5′ phosphoryl ends. Such termini may be 'filled-in' using a DNA polymerizing activity to add nucleotides to the terminal 3′OH groups. This generates blunt ends on the molecules which may then be cyclized using T4 DNA ligase. The 'filling-in' reaction is usually carried out with the Klenow fragment of *E. coli* DNA polymerase I, a proteolytically generated fragment of the enzyme which has the 3′–5′ exonuclease activity but which lacks the 5′–3′ exonuclease. Alternatively the DNA polymerase specified by phage T4 or reverse transcriptase may be used.

An alternative is to insert a chemically synthesized oligonucleotide into a restriction endonuclease cleavage site. This is usually a 'linker' sequence which provides a restriction endonuclease cleavage site marking the position of the insertion. 'Linker' sequences can also be inserted at random positions within the DNA. This is achieved by introducing random double-stranded scissions into covalently closed circular molecules. This requires the action of DNaseI at low concentrations in the presence of manganese ions. The linear DNA that is produced in this reaction is ligated to synthetic linker sequences and then treated with the appropriate restriction endonuclease to expose the cohesive termini of the restriction site. The molecules can then be cyclized using DNA ligase. The position of the insertion mutation can be readily determined since it is marked by a restriction site and can be mapped relative to other restriction endonuclease cleavage sites.

2.3.3 Point mutations

A number of techniques have been used to generate single-stranded regions within DNA in order to apply bisulphite mutagenesis. Bisulphite ions deaminate cytosine residues in single-stranded but not in double-stranded DNA. This treatment changes C residues to U residues and they pair with A residues during DNA replication. Shortle and Nathans [20] have used bisulphite mutagenesis to make mutations in the vicinity of a restriction endonuclease cleavage site. Supercoiled DNA is nicked in one strand by partial digestion with the restriction enzyme in the presence of ethidium bromide. A short single-stranded region is then exposed using the exonucleolytic activity of the 'Klenow' fragment of *E. coli* DNA polymerase I or T4 DNA polymerase in the presence of a single deoxynucleoside triphosphate (as described in Section 2.3.1). This DNA is then treated with sodium bisulphite and the single-stranded DNA is repaired with *pol*I using all four deoxynucleoside triphosphates. This results in the incorporation of an A residue opposite the mutated C residue. Replication *in vivo* then generates an AT base pair (Fig. 2.16). Shortle and coworkers [21] have extended this procedure to mutagenize single-stranded DNA within a D-loop (Fig. 2.15). They incubated D-loops with S1 nuclease under conditions in which only the exposed single-strand was digested, leading to the production of a nicked open circular DNA molecule. The nicks could be converted to gaps using *M. luteus* DNA polymerase I. The 5′–3′ exonuclease associated with this enzyme will degrade five or six bases out from the nick in the absence of nucleoside triphosphates. The gaps can then be mutagenized with bisulphite and the molecules repaired and reintroduced into *E. coli*.

Bisulphite mutagenesis has also been used to introduce point mutations into a region defined by a previously isolated deletion mutant [22]. A plasmid containing the deletion mutant is cleaved with one restriction endonuclease and a plasmid containing the wild type gene is cleaved with a second restriction endonuclease to

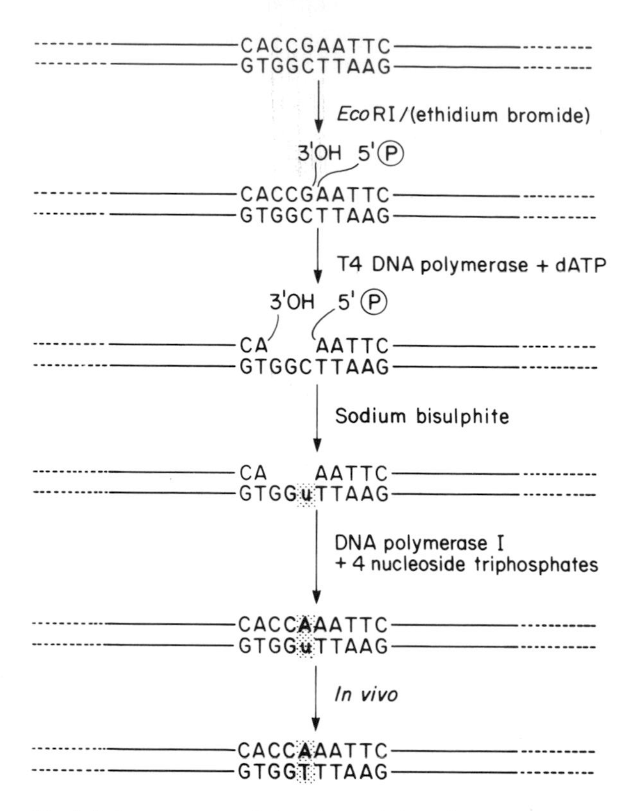

Figure 2.16 Bisulphite mutagenesis in the vicinity of a restriction site.

generate, in both cases, linear molecules cleaved at single sites. The molecules are then denatured and annealed to form circular heteroduplexes in which the wild type strand is 'looped-out' at the position of the deletion (Fig. 2.17). The single-stranded loop of DNA is susceptible to mutagenesis by bisulphite. The mutagenized DNA is introduced directly into an *E. coli* strain that lacks the enzyme uracil-*N*-glycosylase (*ung*$^-$) ensuring that the Uracil residues generated by the bisulphite reaction are not eliminated. Homoduplexes also result from the annealing reaction and will not be mutagenized. These homoduplexes are, however, linear structures and therefore transform *E. coli* at lower efficiency. The majority of the transformants therefore arise from the mutagenized heteroduplex structures. One might expect that the transformant would contain plasmids derived from both strands of the heteroduplex which would segregate upon replication. In practice only one of the two types is recovered as the structure is presumably repaired before replication.

It is also possible to utilize the infidelity of the DNA replicative enzymes in order to make point mutations. Base substitution can be

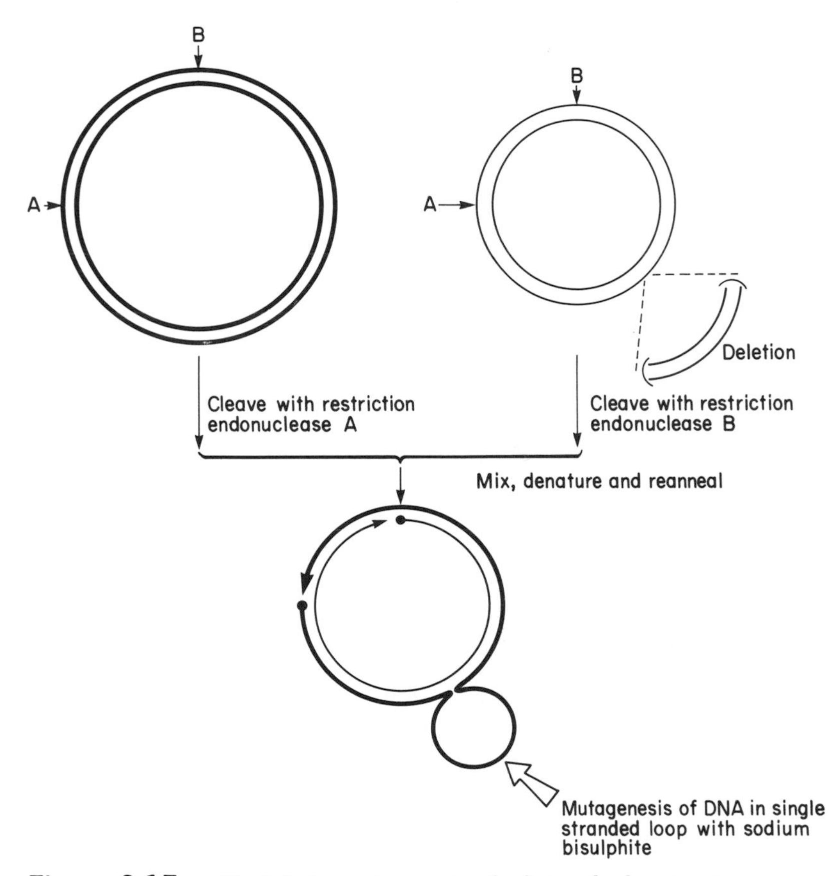

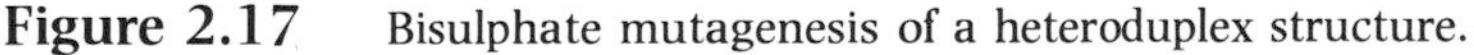

Figure 2.17 Bisulphate mutagenesis of a heteroduplex structure.

carried out by directing the incorporation of nucleotide analogues in the vicinity of restriction endonuclease cleavage sites [23]. A number of restriction enzymes will cleave only one DNA strand within the recognition site, under partial digestion conditions in the presence of ethidium bromide. A base analogue can then be introduced into the DNA in the vicinity of the nick by carrying out the nick-translation reaction catalysed by DNA polymerase I (Fig. 2.18). This reaction, which is routinely used to introduce radiolabelled nucleotides into DNA to make nucleic acid hybridization probes, requires the 5′–3′ exonuclease activity of the enzyme to be in concert with the polymerizing activity which replaces the excised mononucleotides with nucleoside triphosphates supplied to the reaction mixture. In this way dTMP residues can be replaced by hydroxy-dCMP which pairs equally well with A or G. The nucleotide sequence analysis of plasmid DNA mutagenized in this way reveals that AT to GC transitions have taken place.

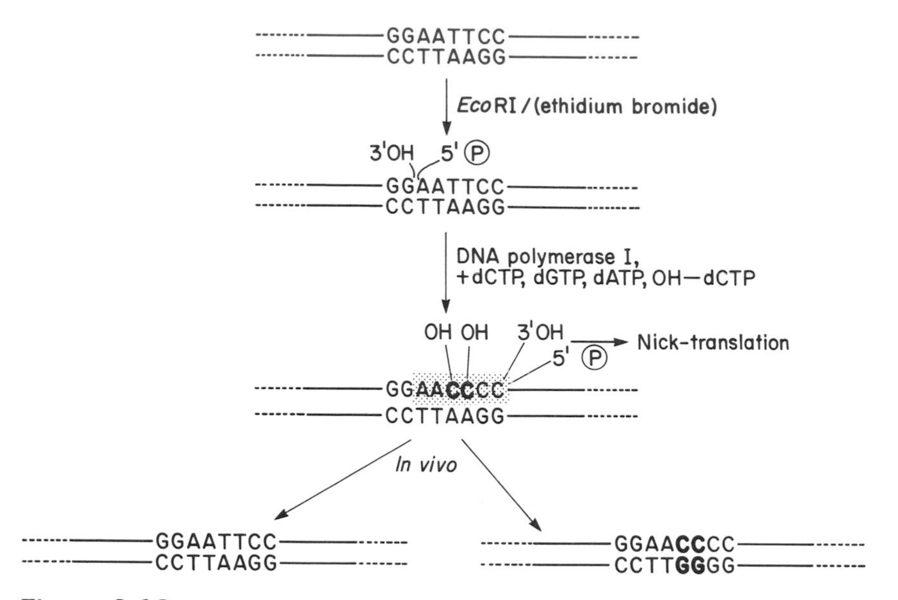

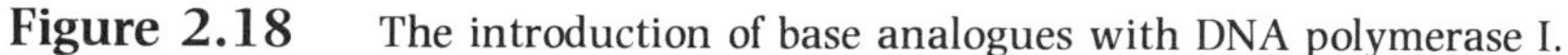

Figure 2.18 The introduction of base analogues with DNA polymerase I.

Other methods have been developed in which DNA polymerase directs the incorporation of an incorrect nucleotide which is not removed by its 3′–5′ exonuclease proof reading mechanism. Shortle *et al.* [24] introduced gaps into plasmid DNA by the exonucleolytic activity of *M. luteus* DNA polymerase I on nicks produced by the action of a restriction endonuclease in the presence of ethidium bromide, as described above. The gaps were filled-in using only three of the nucleoside triphosphates and an excess of DNA ligase. Covalently closed circular DNA molecules were recovered which contained a high proportion of transition or transversion mutations. The problem with this procedure is that misincorporated nucleotides can be removed by the 3′–5′ editing exonuclease. The way around this problem is to incubate the 'gapped' DNA with an α-thionucleoside triphosphate in the absence of other triphosphates. Misincorporated α-thionucleoside triphosphates are not removed by the 3′–5′ exonuclease activity, and so after the subsequent repair of the gap with all four nucleoside triphosphates, mutants can be obtained when the molecules are introduced into *E. coli* (Fig. 2.19).

All of the above techniques for making point mutations rely upon the presence of a restriction endonuclease cleavage site in the region that has to be mutagenized. Techniques are now available to introduce mutations into any site within DNA segments cloned into a single-stranded DNA bacteriophage vector. The method requires the synthesis of an oligonucleotide containing the desired mutation. The oligonucleotide is then annealed with the single-stranded DNA of the

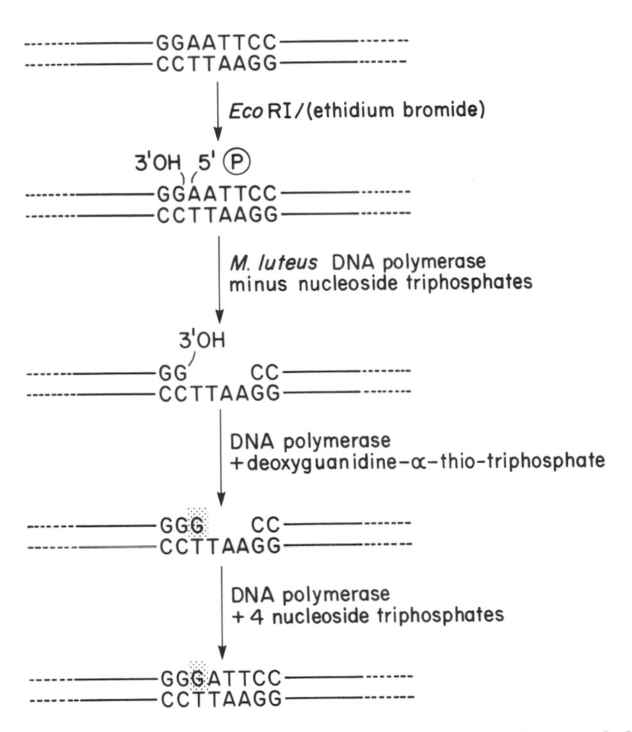

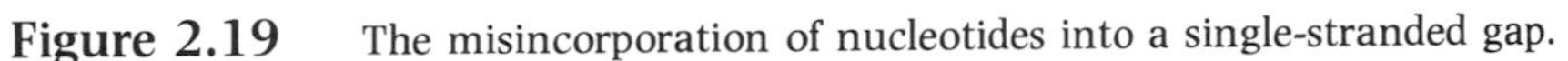

Figure 2.19 The misincorporation of nucleotides into a single-stranded gap.

gene in the vector. A heteroduplex is formed which can be completely repaired using DNA polymerase I and DNA ligase to give covalently closed circular molecules containing mismatched bases in the region corresponding to the oligonucleotide. These are introduced into *E. coli* whereupon the two strands segregate during DNA replication. Progeny phage can be detected by a hybridization screen. The method is described in more detail in Section 3.2.4 as one application of single-stranded DNA bacteriophage vectors.

References

1. Arber, W. (1965) Host controlled modification of bacteriophage. *Ann. Rev. Microbiol*, **19**, 365–78.
2. Meselson, M., Yuan, R. and Heywood, J. (1972) Restriction and modification of DNA. *Ann. Rev. Biochem.*, **41**, 447–66.
3. Roberts, R. J. (1983) Restriction and modification enzymes and their recognition sequences. *Nucleic Acids Res.*, **11**, r135–68.
4. Mertz, J. E. and Davis, R. W. (1972) Cleavage of DNA by R1 restriction endonuclease generates cohesive ends. *Proc. Natn Acad. Sci. USA*, **69**, 3370–4.
5. Lehman, I. R. (1974) DNA ligase: structure, mechanism, and function. *Science*, **186**, 790–7.

6. Sgaramella, V. and Khorana, H. G. (1972) Studies on nucleotides: a further study of the T4 ligase catalysed joining of DNA at base paired ends. *J. Mol. Biol.*, **72**, 493–502.
7. Glover, D. M., White, R. L., Finnegan, D. J. and Hogness, D. S. (1975) Characterisation of six cloned DNAs from *Drosophila melanogaster*, including one that contains the genes for rRNA. *Cell*, **5**, 149–57.
8. Maniatis, T., Hardison, R. C., Lacy, E. *et al.* The isolation of structural genes from libraries of eukaryotic DNA. *Cell*, **15**, 687–701.
9. Lobban, P. and Kaiser, A. D. (1973) Enzymatic end to end joining of DNA molecules. *J. Mol. Biol.*, **78**, 453–71.
10. Jackson, D. A., Symons, R. M. and Berg, P. (1972) Biochemical method for inserting new genetic information into DNA of Simian virus 40: circular SV40 DNA molecules containing λ phage genes and the galactose operon of *Escherichia coli*. *Proc. Natn Acad. Sci USA*, **69**, 2904–9.
11. Wensink, P. C., Finnegan, D. J., Donnelson, J. E. and Hogness, D. S. (1974) *A system for mapping DNA sequences in the chromosomes of Drosophila melanogaster*. *Cell*, **3**, 315–25.
12. Roychoudhury, R., Jay, E. and Wu, R. (1976) Terminal labelling and addition of homopolymer tracts to duplex DNA fragments by terminal deoxynucleotidyl transferase. *Nucl. Acids. Res.*, **3**, 101–16.
13. Itakura, K. (1982) Chemical synthesis of genes. *Trends in Biochem. Sci.*, **7**, 442–5.
14. Maniatis, T., Kee, S. E., Efstratiadis, A. and Kafatos, F. (1976) The amplification and characterisation of a β-globin gene synthesised *in vitro*. *Cell*, **8**, 163–82.
15. Okayama, H. and Berg, P. (1982) High efficiency cloning of full length cDNA. *Mol., Cell Biol.*, **2**, 161–70.
16. Carbon, J., Shenk, T. E. and Berg, P. (1975) Biochemical procedure for production of small deletions in Simian virus 40 DNA. *Proc. Natn Acad. Sci. USA*, **72**, 1392–6.
17. Gray, H. B., Ostrander, D. A., Hodnett, J. L., Legerski, R. J. and Robberson, D. L. (1975) Extracellular nucleases of *Pseudomonas* BAL31: characterisation of single strand specific deoxyriboendonuclease and double strand deoxyriboexonuclease activities. *Nucl. Acids Res.*, **2**, 1459–92.
18. Panayotatos, N. and Truong, K. (1981) Specific deletion of DNA sequences between preselected bases. *Nucl. Acids Res.*, **9**, 5679–88.
19. Green, C. and Tibbetts, C. (1980) Targetted deletions of sequences from closed circular DNA. *Proc. Natn Acad. Sci. USA*, **77**, 2455–9.
20. Shortle, D. and Nathans, D. (1976) Local mutagenesis: a method for generating viral mutants with base substitutions within preselected regions of the viral genome. *Proc. Natn Acad. Sci. USA*, **75**, 2170–4.
21. Shortle, D., Koshland, D., Weinstock, G. M. and Botstein, D. (1980) Segment directed mutagenesis: construction *in vitro* of point mutations linked to a small predetermined region of a circular DNA molecule. *Proc. Natn Acad. Sci. USA*, **77**, 5375–9.
22. Kalderon, D. Oostra, B. A., Ely, B. K. and Smith, A. E. (1982) Deletion loop mutagenesis. A novel method for the construction of point mutations using deletion mutants. *Nucl. Acids. Res.*, **10**, 5161–71.

23. Muller, W., Weber, H., Meyer, F. and Weissmann, C. (1978) Site directed mutagenesis in DNA: generation of point mutations in cloned β-globin complementary DNA at the positions corresponding to amino acids 121 to 123. *J. Mol. Biol.*, **124**, 343–58.
24. Shortle, D., Grisafi, P., Benkovic, S. J. and Botstein, D. (1982) Gap misrepair mutagenesis: efficient site directed induction of transition, transversion, and frame shift mutations *in vitro*. *Proc. Natn Acad. Sci. USA*, **79**, 1588–92.

3 Bacteriophage vectors

3.1 Bacteriophage λ

Of all the bacteriophage which infect *E. coli*, bacteriophage λ has been the most extensively studied. The wealth of knowledge of the molecular biology and genetics of the phage [1, 2] has greatly assisted its development as a cloning vehicle. It has about 50 genes in its 49 kb genome and of these only about half are essential. The non-essential genes can be replaced with foreign DNA and the recombinant genome DNA can then be propagated in *E. coli*. This chapter will trace the development of a versatile series of vectors for different cloning problems.

3.1.1 The biology of phage λ

The infection of *E. coli* with phage λ is initiated by absorption of the phage onto the bacterium. This involves an interaction between the tip of the phage tail with a component of the outer cell membrane of *E. coli*. Following penetration, the linear double-stranded DNA molecule cyclizes through the *cos* sites at its ends. These termini are single-stranded, mutually complementary sequences of 12 residues in length [3]. Early in the infection cycle these circular DNA molecules replicate as theta (θ) forms. The replication is bidirectional [4]; it originates between the genes *cII* and *O* and requires the activity of the phage genes *O* and *P*. Later in infection there is a switch towards a rolling circle mechanism which produces long concatameric molecules composed of several linearly arranged genomes [5]. The switch is brought about by the action of the phage *gam* gene product. This inhibits the action of the *E. coli recBC* nuclease upon the replicating DNA [6]. *Gam*$^-$ phages can therefore only produce concatameric molecules in *rec*BC$^-$ hosts. Since multimers of the phage genome are needed for packaging into virions, the maturation of *gam*$^-$ phage in *rec*BC$^+$ hosts requires recombination to occur (Fig. 3.1). This is brought about by either the phage *red* gene or the host *rec*A gene product. The multimeric λ genomes which are products of recombination, or the concatomeric genomes resulting from rolling circle replication, are cleaved to unit length genomes by the action of a nuclease at the *cos* sites [7]. This process, which also requires the presence of four head proteins, will be discussed later in this chapter (Section 3.1.3(a)).

The bacteriophage genome has an alternative mode of propagation whereby it becomes stably integrated into the host chromosome and is replicated along with the bacterial chromosome. In this state only

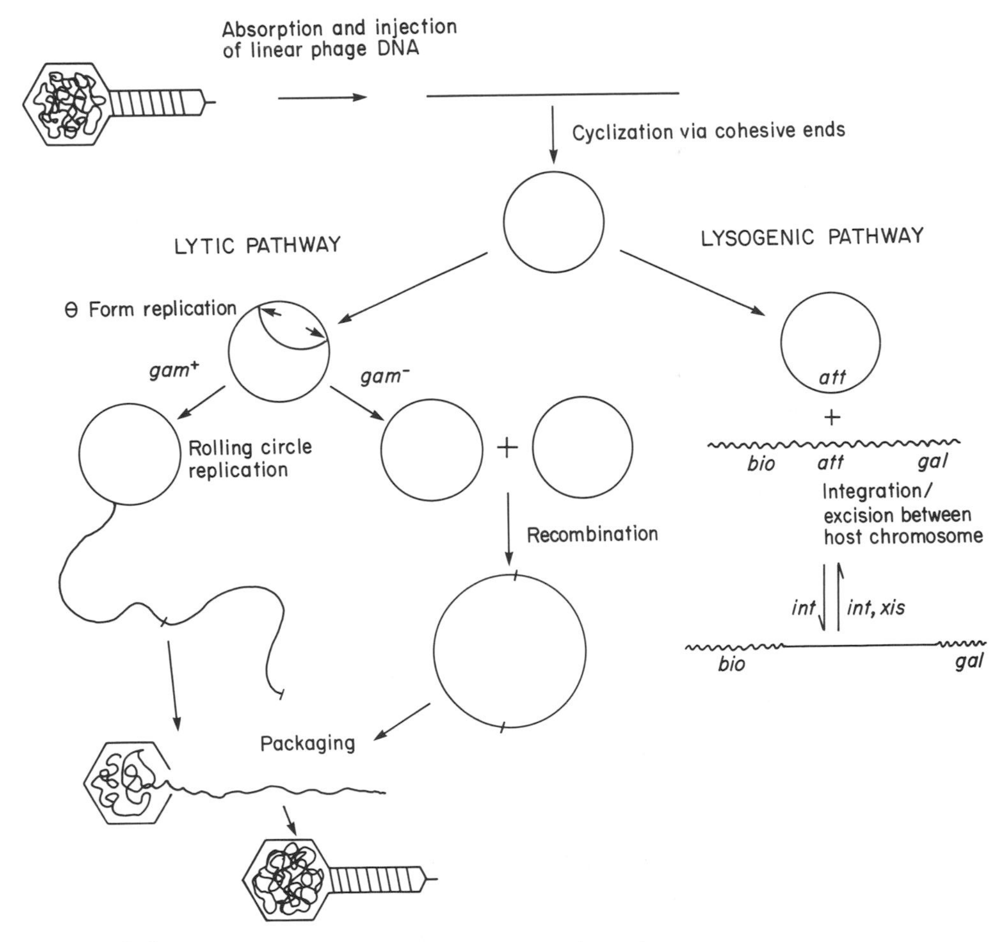

Figure 3.1 The replication cycles of bacteriophage λ.

the *cI* gene is expressed to produce a protein which represses all of the genes responsible for the lytic pathway. The integration of the phage genome occurs by recombination between the phage attachment site (*att*) and a partially homologous site on the *E. coli* chromosome. Integration, which requires expression of the phage gene *int*, is a reversible process. Prophage excision requires activity of a phage gene *xis* in concert with *int*. The normal *att* site maps between the genes for galactose utilization (*gal*) and biotin biosynthesis (*bio*) (Fig. 3.1). Abnormal excisions of the prophage can result in the incorporation of genes from one or other of these operons into the phage genome, with the concomitant deletion of some phage DNA. Depending upon the extent of this deletion, these transducing phage may or may not be defective for vegetative growth or lysogenization. If the normal attachment site is deleted from the *E. coli* chromosome then the phage can integrate at secondary attachment sites. Abnormal excisions of prophage from these other regions of the genome can result in the transducing phage carrying a number of other *E. coli* genes.

The decision between the lytic and lysogenic events is dependent upon the interactions of two proteins, the *c*I gene product and the *cro* gene product, with the two λ operators. The following is a highly simplified account of the sequence of gene expression for the two responses and serves to introduce the reader to various promoters in the genomes, some of which can be used to enhance the expression of cloned DNA sequences. If the two operators are free of repression then leftward and rightward transcription can take place on opposite strands from the two promoters p_L and p_R. Leftward transcription terminates at t_L to give 12S mRNA which encodes the *N* gene product. The majority of the rightward transcripts terminate at t_{R1} to give the 7S mRNA for the *cro* gene product [8]. There is some readthrough to t_{R2} to give transcripts of the *cII*, *O* and *P* genes. Readthrough at each of these three terminators is enhanced by the *N* gene product [9, 10] and this leads to synthesis of the *cIII* gene product and those gene products involved in recombination from the leftward transcription unit. It also results in elevated levels of the *cII*, *O* and *P* gene products and also of the *Q* gene product from rightward transcription. At this stage the *cro* gene product has reached such a concentration that it will shut down transcription from p_L and p_R by binding to the operators. The *Q* gene product meanwhile activates very efficient transcription from p'_R to ensure large quantities of head and tail proteins (Fig. 3.2).

The products of the *cII* and *cIII* genes can stimulate transcription from p_{re} the promoter for the establishment of lysogeny resulting in the synthesis of the *cI* gene product. This binds tightly to p_L and p_R and if sufficient repressor is made before late transcription can

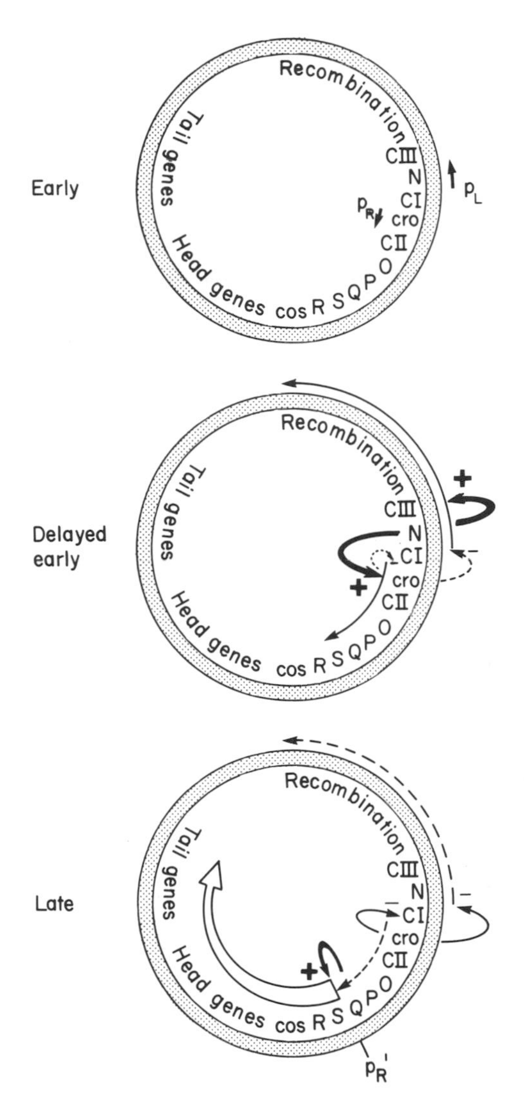

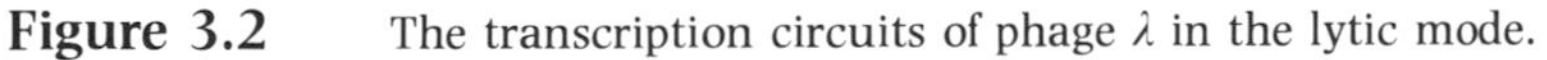
Figure 3.2 The transcription circuits of phage λ in the lytic mode.

become established then all the genes of the phage will be repressed and the genome will be integrated into the chromosome by the *int* gene product. The repression of the *cII* and *cIII* genes in turn leads to the cessation of transcription from p_{re}. The leftward and rightward operators each contain three binding sites for the *cI* repressor and all of these are filled when the repressor concentration is high. Upon division of the lysogenized bacterium, the sites empty in order of their binding affinity. When only one of these sites is bound to repressor the p_{rm} promoter (repression maintenance) is activated and more *cI* repressor can be synthesized [11] (Fig. 3.3). By incorporating a temperature-sensitive mutation in the *cI* gene, such that the polypeptide is inactivated at high temperature (usually 42–45° C), it is possible to construct lysogens which can be conveniently induced by shifting the temperature of the culture. In nature, induction usually occurs in response to a mutagen such as ultraviolet light, which is thought to act by inducing the host *rec*A protein which can proteolytically cleave and so inactivate the *cI* repressor [12]. This allows the initiation of leftward and rightward transcription from p_L and p_R. Phage λ is one member of the family of lambdoid phages, all of which are active in *E. coli*, but which display different immunities [13]. Immunity to superinfection is a characteristic of lysogenic cells resulting from the presence of *cI* repressor in their cytoplasm. The λ *cI* gene product will not, however, repress phage 434 or phage 21, for example, since the operators of these phages will not bind the λ repressor. These phage genomes, however, do have considerable regions of sequence homology and so it is possible to construct recombinant λ phage which carry the immunity region (the operators, *cI* and *cro* genes) of phage 434 for example. Such a phage is called λimm434 and we shall see that such hybrid phage provide a useful means of varying the number of restriction sites for a given enzyme in phage λ vectors.

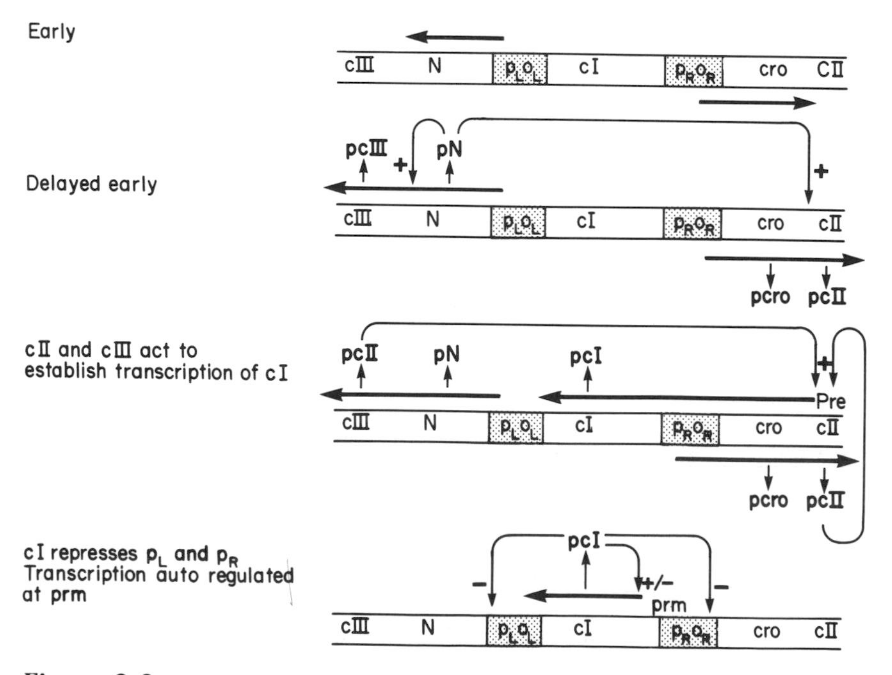

Figure 3.3 Lambda promoters for the establishment and maintenance of lysogeny.

3.1.2 Phage vectors

(*a*) *Eliminating restriction sites*

Before phage λ could be used as a cloning vehicle, it was first necessary to eliminate from its genome some of the restriction sites for the enzymes commonly used for cloning. Wild type λ has five cleavage sites for *Eco*RI. These are displayed on the linear map of the genome in Fig. 3.4. Derivatives of λ were constructed having a reduced number of *Eco*RI sites. The remaining *Eco*RI sites are in a non-essential region of the genome so that phage DNA might be cleaved to completion with *Eco*RI and foreign DNA inserted into this

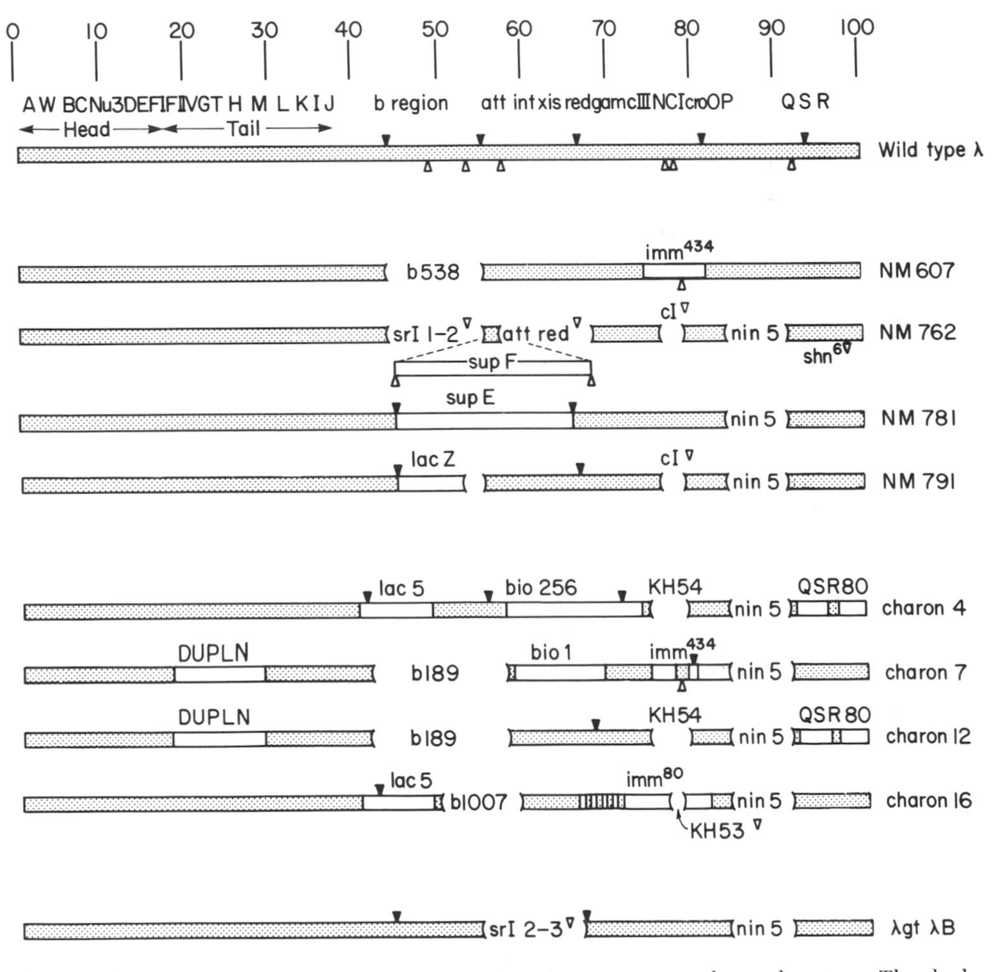

Figure 3.4 Physical maps of wild type λ and some commonly used vectors. The dark triangles above the map represent *Eco*RI sites and the open triangles below the map represent *Hin*dIII sites.

region. The strategy was first to remove all the sites and then to replace the desired ones by *in vivo* genetic recombination. Rambach and Tiollais [14] and Murray and Murray [15] began such experiments with phage which already had naturally occuring deletions and which did not have the two left most sites (sites 1 and 2). Thomas *et al.* [16] used a transducing phage which had also lost site three as a result of a *bio* substitution. This phage also had a sequence duplication in its left arm to facilitate its propagation. This is because if the length of the phage genome falls below 75% of the wild type length the DNA is no longer packaged into the capsids. A corresponding packaging limitation occurs if the length of a phage molecule exceeds the wild type length by 10%. These factors are clearly important to bear in mind both when designing phage vectors *per se* and also because of the limitations they impose upon the lengths of recombinant phage DNAs. Mutants lacking the remaining sites were selected by cycling the phage between hosts containing, and hosts lacking, the *Eco*RI restriction–modification system. Restriction modification systems were encountered in Chapter 2. In an infection of a host carrying the *Eco*RI restriction–modification system by unmodified phage, the progeny phage are vastly reduced in number and have arisen from DNA molecules that were modified early in infection and so protected from the restriction endonuclease. They may in addition have mutations at *Eco*RI sites such that they can no longer be cleaved by the restriction enzyme. By repeatedly cycling the phage between hosts, phage are selected that have acquired several such mutations. The cycling is continued until the efficiency of plating indicates that the DNA can no longer be restricted. The resulting phage which completely lack *Eco*RI sites are then crossed *in vivo* with phage containing all the *Eco*RI sites, and recombinants selected which contain only sites 1 and 2, or 1, 2 and 3. These phage can be cleaved with *Eco*RI and fragments of foreign DNA inserted between the left and right phage arms.

Phage λ vectors have also been constructed for the restriction fragments produced by *Hin*dIII [17]. The wild type phage DNA has six sites for this enzyme which are also shown on the map in Fig. 3.4. It can be seen from Fig. 3.4 that deletion of the restriction fragment between *Eco*RI sites 1 and 2 removes two *Hin*dIII sites. Also phage having the b538 deletion lack the *Hin*dIII sites 1, 2 and 3. The substitution of the immunity region for phage 21 (*imm*^{21}) for *imm*λ removes *Hin*dIII sites 4 and 5. Site 6 is lost by the substitution of ϕ80 DNA in this region of the genome. This has been achieved by selecting for *in vivo* recombinants around the *Q* gene.

The vectors which have been developed using the basic principles described above fall into two categories: those with a single restriction site into which one can insert foreign DNA; and those with two restriction sites that allow a non-essential segment of vector DNA to be replaced with foreign DNA. The packaging limitations place restrictions upon the lower and upper size for a recombinant λ molecule. In practical terms therefore the insertion vectors can accommodate less DNA and are better suited to tasks such as cDNA cloning or the cloning of small restriction fragments. On the other hand the replacement vectors can accommodate larger segments and are better suited to cloning regions of chromosomal DNA from the higher eukaryotes where the genes are physically larger because of the numerous intervening sequences they contain.

(*b*) *Insertion vectors*

The immunity region of other lambdoid phages has been introduced into bacteriophage λ to provide cloning sites in hybrid phage vectors. The immunity region of phage 434 contains single cleavage sites for *Hin*dIII and *Eco*RI. This region has been incorporated into the genome as a result of crosses between the two phage. There is another advantage in having cloning sites within the *cI* gene. In a cloning experiment the viable products of the ligation reaction are either recombinants (provided the inserted DNA does not cause the genome to exceed the packaging limit), or molecules in which the left and right arms have rejoined to generate the parental phage. The insertion of foreign DNA inactivates the *cI* genes, resulting in clear plaque morphology. The *in vitro* recombinant phage are thereby distinguished from parental vector phage which can lysogenize and so give turbid plaques (Fig. 3.5). Some phages from this series of λimm^{434} hybrid phage vectors are illustrated in Fig. 3.4 [17]. Similar insertion vectors are to be found in the 'Charon' series developed by Blattner *et al.* [18] (Fig. 3.4). Charon 6 and Charon 7 also contain the imm^{434} region. Charon 7 will serve as an insertion vector for both *Hin*dIII and *Eco*RI, whereas Charon 6 only serves as an effective insertion vector for *Eco*RI fragments, since it retains a second *Hin*dIII site (site 6). There are alternative insertion vectors in the Charon series: insertion into the single *Eco*RI site of the *red* gene of Charon 12 should make the phage unable to grow on *E. coli* with mutant DNA polymerase I. Similarly, insertion of DNA into the single *Eco*RI site of Charons 2 or 16 should inactivate the *lacZ* gene carried on these phage. Active β-galactosidase can be detected by plating cells on media containing the *lac* operon inducer IPTG and the chromogenic substrate for the enzyme 4-bromo-5-chloroindol-3-ylβ-D-galactoside (XG). The active enzyme causes the formation of blue colonies or plaques whereas if the enzyme is inactive they remain white (see also below).

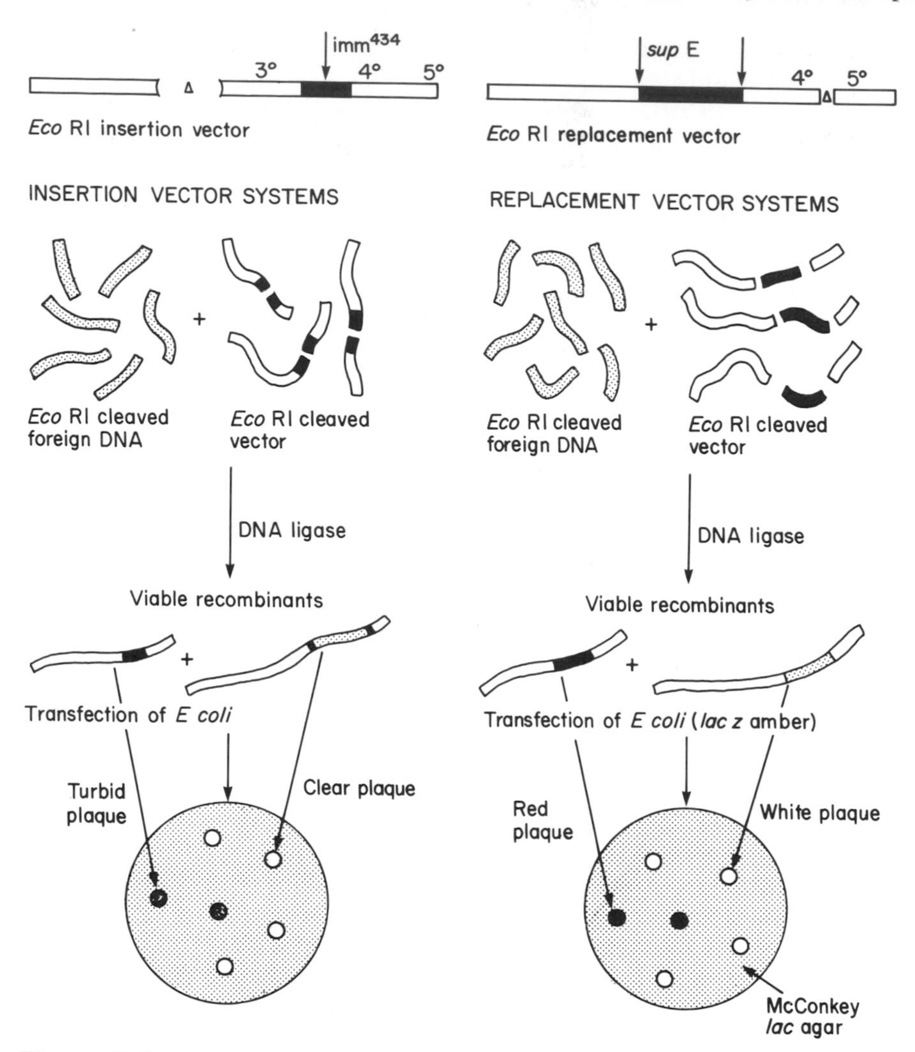

Figure 3.5 Insertion vectors and replacement vectors.

A direct selection can be applied to vectors in which the *cI* gene is inactivated due to the insertion of foreign DNA. This utilizes a mutation in the host gene *hfl*A. The wild type gene product of *hfl*A appears to act by suppressing the synthesis of the phage *cII* gene product [19]. The establishment of lysogeny requires the coordinate production of *cI* repressor and integrase. The transcription of both of these genes is positively regulated by the *cII* gene product (Fig. 3.3). Mutations in the *E. coli* gene *hfl* result in a high concentration of the *cII* gene product and this leads to the the establishment of lysogeny. The parental insertion vector plated on a *hfl*A strain does not undergo the lytic infection cycle, but forms lysogens. Recombinants in which the *cI* gene is inactivated, on the other hand, form plaques. This forms the basis of a selection which has been employed in amplifying cDNA libraries of developmentally regulated *Drosophila* genes cloned in the *cI* gene of the insertion vector λ641 [20].

Another useful insertion vector for cDNA cloning has been described by Young and Davis [21]. This phage, λgt11, carries the *lacZ* gene from *E. coli* which has a single *Eco*RI site 53 base pairs upstream from the termination codon. It can accommodate 8.3 kb of foreign DNA and is designed as an expression vector to facilitate the isolation of specific genes by the antibody screening technique described in Chapter 1. The insertion of DNA inactivates the *lacZ* gene, making the phage unable to produce blue plaques on a *lacZ*$^-$ host on plates containing X-gal. If the insertion maintains the correct orientation and translational reading frame with the *lacZ* gene then the recombinant phage can direct the synthesis of fusion proteins between β-galactosidase and the peptide specified by the foreign DNA. One sixth of the recombinants should therefore produce peptides of interest. The bacterial host used in conjunction with this vector contains two useful mutations (Fig. 3.6). The first is in the *lon*$^-$ mutation which causes a defect in protein degradation pathways. It has been noticed that the stability of a plasmid-encoded hybrid protein containing the sequence of the hormone somatostatin is enhanced in such hosts (see also Chapter 5, Section 5.2). The second mutation is in the *hfl*A gene. This ensures that infected cells produce lysogens rather than undergoing the lytic cycle. Once a phage library has been made in λ-gt11 it can be introduced into the *E. coli hf1* A to produce lysogens which can be screened using antibodies. In order to permit efficient screening the phage genome carries the *cI*857 gene so that the lysogens can be induced by shifting the temperature from 32–42° C. An amber mutation in the *S* gene (see Section 3.1.3) renders the phage deficient for cell lysis and consequently the cells fill the large amounts of phage products. The thermo-induced colonies can be lysed by exposure to chloroform vapour and the antibody screening carried out as described in Chapter 1. Kemp and his colleagues have modified λ-gt11 to incorporate the *β-lactamase* gene from the plasmid pBR322 (see Chapter 4). This specifies resistance to ampicillin and therefore provides a means for the direct selection of lysogenic colonies. They have used such a vector to construct a cDNA library from the malarial parasite *Plasmodium falciparum* [22]. Recombinant phage have been identified that express antigens recognized by immune human sera capable of inhibiting the growth of *P. falciparum*. These cloned antigens should facilitate the production of potentially useful vaccines.

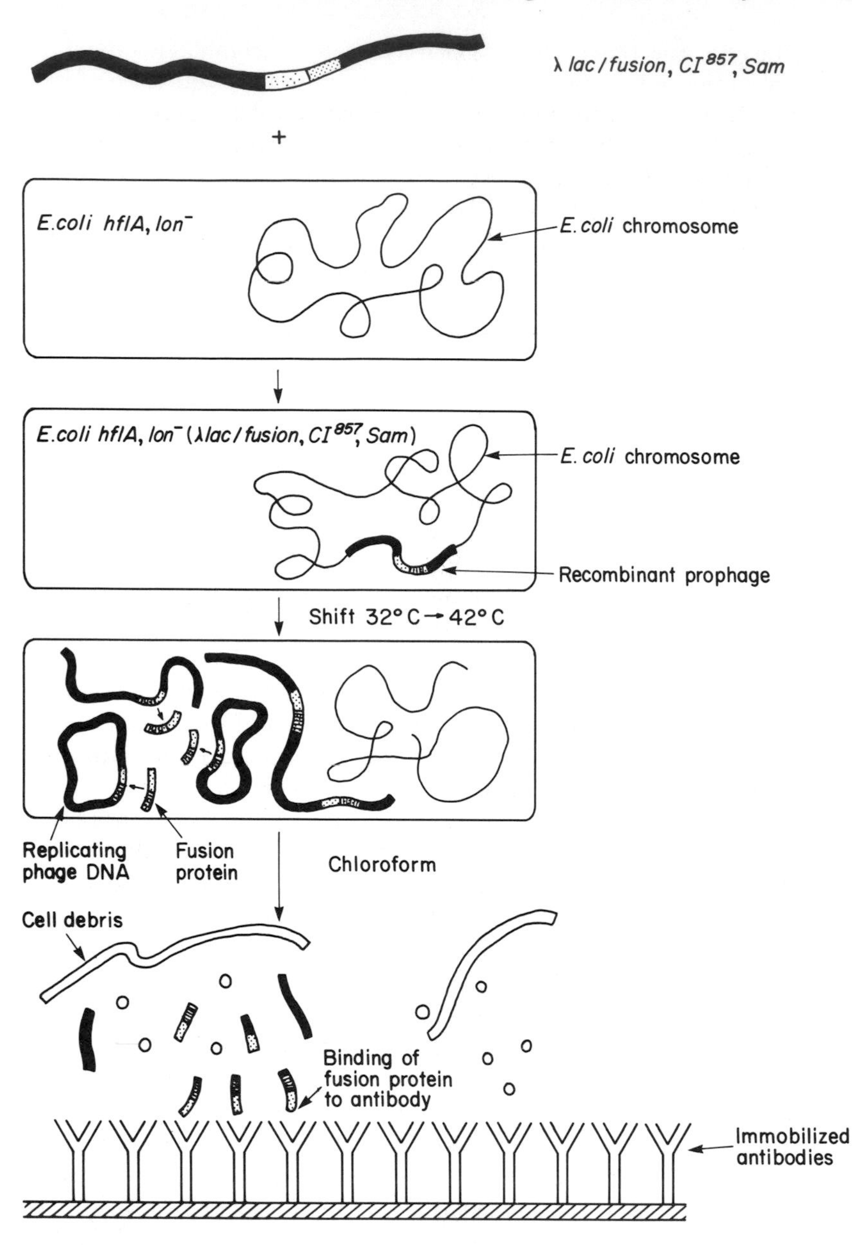

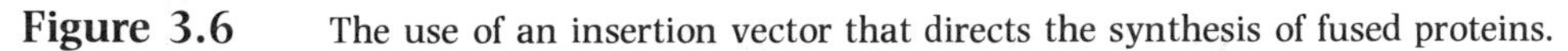

Figure 3.6 The use of an insertion vector that directs the synthesis of fused proteins.

(*c*) *Replacement vectors*

The formation of plaques by *in vitro* recombinant phage can be made to depend upon the reconstitution of a molecule of a certain size, since the DNA will only be packaged into mature phage if its length is within the packaging limits. The amount of DNA that can be cloned in phage λ can therefore be increased by using a phage that has two sites for a restriction endonuclease bordering a non-essential region. The non-essential phage DNA serves to assist propagation of the vector and is subsequently replaced with foreign DNA. The packaging limitations ensure that the viable products of *in vitro* recombination must either have had this central fragment reincorporated into the genome or replaced by foreign DNA. Deletion of this fragment would generate a DNA molecule too small to be packaged.

In the replacement vector, λgt.λC, of Thomas *et al.* [16], the replacement fragment lies between *Eco*RI sites 2 and 3 (fragment C). It contains *att*, *int* and *xis* and gives the phage the capability of forming stable lysogens. When fragment C is replaced by foreign DNA, however, the phage becomes integration defective. Murray *et al.* [17] have described a series of replacement vectors for *Eco*RI or *Hin*dIII fragments which simplify the recognition of recombinants. One such phage, λNM781, has a replaceable *Eco*RI fragment which carries the gene, *sup*E, for a mutant tRNA of *E. coli* (Fig. 3.4). This gene is active when the *Eco*RI fragment is oriented in either direction in the chromosome. The phage is recognized by the suppression of an amber mutation in the *lac*Z gene of the bacterial host, either as red plaques on lactose/MacConkey agar or as blue plaques on agar containing X-gal. The *in vitro* recombinant phage give colourless plaques on both these indicators (see Fig. 3.5). A similar replacement vector for *Hin*dIII fragments is described, NM762, containing a *Hin*dIII fragment for the *sup*F gene.

An alternative vector (λNM791) has most of the *lac*Z gene as a replaceable *Eco*RI fragment which can be recognized by allelic complementation of a suitable *lac*$^-$ indicator strain. Detection methods of this type are also used with M13 vectors and are discussed in detail in Section 3.2.2. Many Charon phages also carry the *lac*5 substitution which contains the β-galactosidase gene together with its operator and promoter. In the replacement vector Charon 4, for example, most of the *lac* DNA is on a single *Eco*RI fragment and this, together with the adjacent *Eco*RI fragment which contains a *bio* substitution, can be replaced by foreign DNA. The parental phage will give dark blue plaques when plated on medium containing the β-galactosidase substrate, X-gal. When the *Eco*RI fragment containing *lac*5 is replaced with foreign DNA, colourless plaques are produced. If the fragment is rearranged in Charon 4 as a result of *in* *vitro* recombination, or if in the case of Charon 16 a foreign sequence is inserted into the *lac* gene, then the plaques will be colourless on *lac*$^-$ indicator bacteria but pale blue on *lac*$^+$ strains. In the latter case, the increased gene dosage of *lac* operator produced by phage growth is thought to titrate out the cell's *lac* repressor, so causing some derepression of the bacterial *lac* operon [18].

(d) Positive selection with the Spi$^-$ phenotype

Several replacement vectors have been designed in which the central restriction fragment carries the *red* and *gam* genes of phage λ. The loss of the *red* and *gam* genes confers the *Spi*$^-$ phenotype upon the phage, so enabling it to form plaques on strains of *E. coli* lysogenic for phage P2. Wild type phage will not form plaques on such lysogens and are *Spi*$^+$ (sensitive to P2 interference). If the phage *red* and *gam* genes are replaced with foreign DNA, the recombinant phage genomes can be selected directly from the products of a ligation mixture by plating on a P2 lysogen. It was briefly discussed earlier in this chapter that *red*$^-$ *gam*$^-$ phages cannot be propagated in *rec*A$^-$ hosts. Indeed their propagation in *rec*$^+$ hosts requires that the phage contains a *chi* site, a sequence needed for recombination. The phenotypes of both wild type and *red*$^-$ *gam*$^-$ phages on various hosts are summarized in Table 3.1

Table 3.1 The plaque phenotypes of wild type and *red*$^-$ *gam*$^-$ phages

	+	*rec*A$^-$	*rec*A$^-$ *rec*BC$^-$	P2 lysogen
λ *red*$^-$, *gam*$^-$	very small	–	+	minute
λ *red*$^-$, *gam*$^-$, χ	small	–	+	small
λ^+	+	+	+	–

We have already encountered several phage which become *red*$^-$ *gam*$^-$ as recombinant molecules. Several of these phages, for example Charon 4, do not contain *chi* sequences and therefore do not grow particularly well on many hosts. *Chi* sequences may be fortuitously provided by the foreign DNA in such recombinants, in which case these phage would have a growth advantage within the population that makes up a library.

A number of λ vectors have now been constructed which allow the *Spi* selection and which contain *chi* sites. Loenen and Brammar [23] have constructed a phage, L47, which can be used as a replacement vector for fragments generated by *Bam*HI, *Eco*RI or *Hin*dIII. The recombinants may be selected by their *Spi*$^-$ phenotype and the left phage arm contains a *chi* site to facilitate propagation of the recombinant. A second such vector, λ1059 has been constructed specifically as a vector for restriction fragments which can anneal to the *Bam*HI cohesive end [24]. λ1059 is a so called 'phasmid' vector. It carries a ColEl plasmid having cloned λ*att* sites. Such recombinants can be grown lytically as phages or as plasmids in the presence of λ repressor. The plasmids can be released from the phage arms by infecting *E. coli* that is constitutively producing λ integrase. One problem with such a vector is that a minor proportion of parental phage do survive the *Spi*$^-$ selection, and these are subsequently picked up in screens for foreign genes by nucleic acid hybridization (see Chapter 1). This is because such screens are almost invariably carried out with probes made from recombinants in which the plasmid vectors have homology to ColEl. This problem has been circumvented in λEMBL1 in which the *Hin*dIII fragment of λ1059 which contains the plasmid sequence has been replaced by a fragment carrying the *E. coli trp*E gene [25]. Variants of λEMBL1 have been selected which have lost the *Eco*RI sites (see Fig. 3.7) and the *Bam*HI sites replaced by linkers having the recognition sequences for *Eco*RI: *Bam*HI: *Sal*I. The phages λEMBL3 and λEMBL4 contain this linker in opposite orientations. They permit the cloning of large fragments (9–22 kb) generated by any of these enzymes. Furthermore they allow an approach to the cloning of randomly

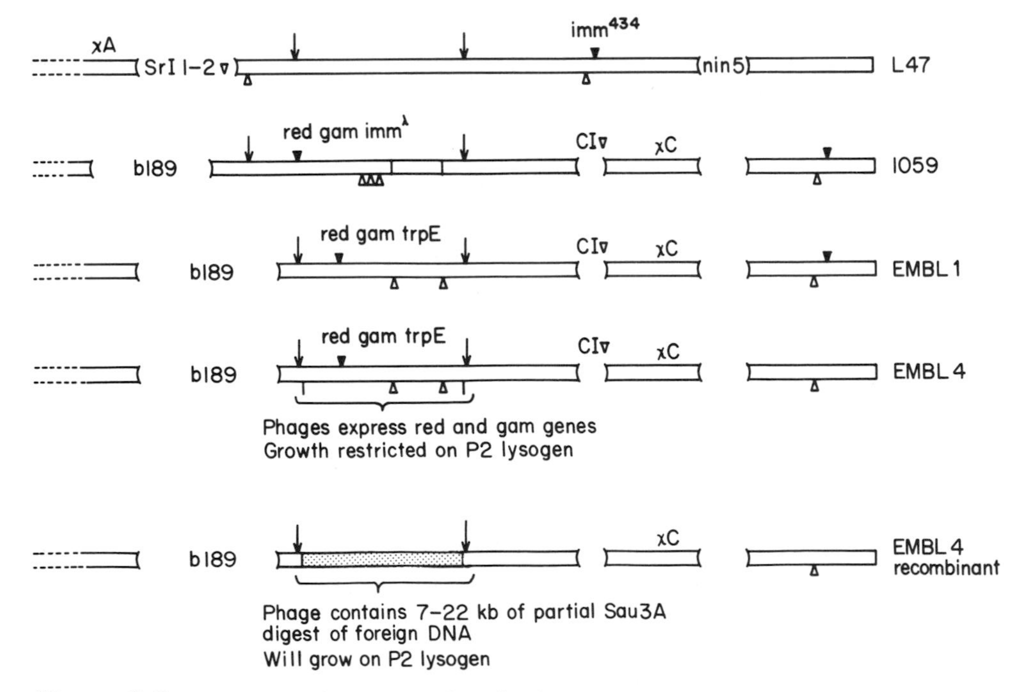

Figure 3.7 Vectors that permit *Spi* selection.

fragmented chromosomal DNA of complex genomes that is greatly simplified compared with the approach discussed in Chapter 2 (see Fig. 2.3). Chromosomal DNA can be partially digested with *Sau*3A which recognizes the sequence GATC and so generates an identical cohesive terminus to that made by *Bam*HI. Such partially digested DNA can therefore be size fractionated and cloned into the *Bam*HI site of λEMBL3 and λEMBL4. The *Bam*HI site is destroyed in this reaction, but the foreign DNA may still be recovered by digesting with either *Eco*RI or *Sal*I according to the vector that was used.

3.1.3 The late genes – their exploitation in cloning vectors

(a) *Phage assembly*

The genes for the late proteins are effectively transcribed from p'_R, providing it has been activated by the *Q* gene product. The late genes include *S* and *R*, which encode proteins required for host lysis and some 20 genes for head and tail proteins. *In vitro* complementation studies [26, 27] have played an important part in elucidating the process of phage maturation. Head and tail assembly occur separately, so that phage with mutations in head genes can only make tails, and phage with mutations in tail genes can only make heads. A mixture of extracts from these two classes of mutant enables viable phage particles to be assembled *in vitro*. This assay has been extended to encompass complementation in pairs of head mutants and has proved invaluable in establishing the sequence in which the phage maturation functions act. The products of genes *E* and *D* (pE and pD) are the major capsid proteins which account for 72% and 20% of the total head protein respectively. The other minor head components are: pB*, derived by proteolytic cleavage of the gene *B* product; X1 and X2, two similar polypeptides derived from the fusion of part of pE with pC; pF11 and pW, which are needed for the head to combine with a tail. A model for the sequential process of head assembly is shown in Fig. 3.8. Small empty headed particles known as *petit lambda* (pλ), are produced in normal infections although differing forms of pλ are produced in infections with either C^- or B^-C^- phage and in phage grown in *E. coli* with *gro*E mutation. These appear to contain a protein core which is lost in the transition to the prohead II structure. The cleavage of pB to pB* and the cleavage and fusion of pC and pE to give X1 and X2 require the putative core protein pNu3. The products of genes *A* and *D* then act to initiate filling of the shell with DNA and the expansion of the shell to its mature size. The concatameric DNA molecules produced late in infection are cleaved to unit length by the action of pA at the *cos* site.

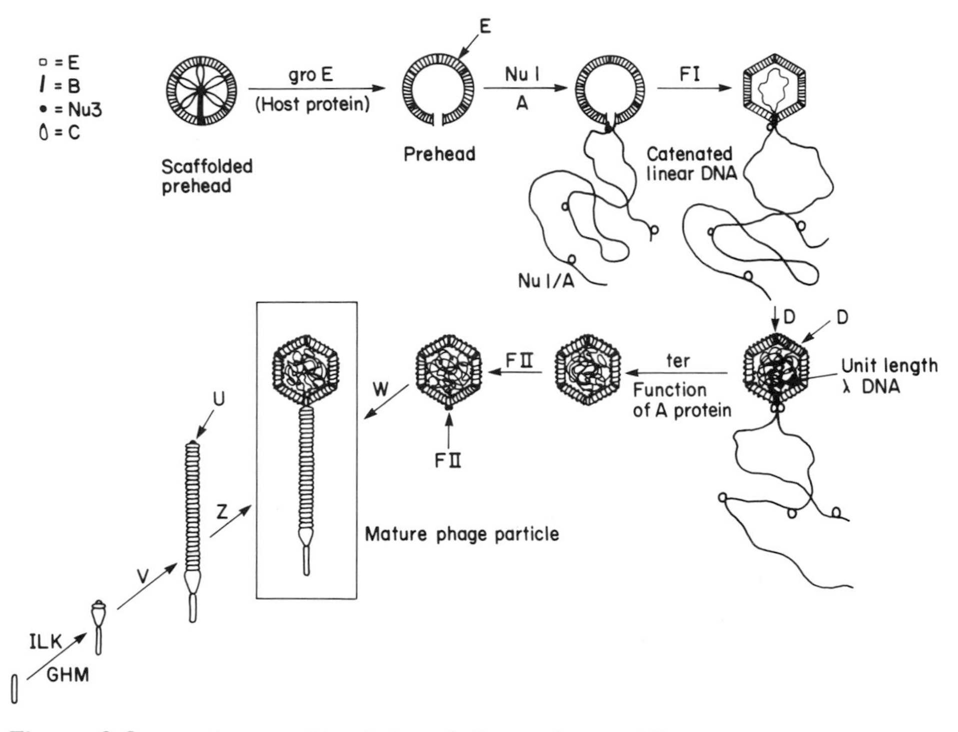

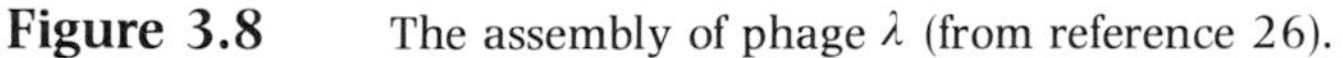
Figure 3.8 The assembly of phage λ (from reference 26).

Finally pW and pF11 are added in sequence to generate a structure that is able to bind mature tails [26, 27].

(*b*) *Biological containment*

Early work with recombinant DNA molecules was beset with difficulties because of the worry that a potentially hazardous gene could be inadvertently cloned and expressed within *E. coli*. In order to provide λ vectors with a level of biological containment, amber mutations have been introduced into a number of the late genes. These phages can then be propagated only on a host containing an amber suppressor mutation. It is therefore highly unlikely that they would be able to grow in the wild type *E. coli* that populate the gastrointestinal tract. The potential risk from the accidental ingestion of recombinant DNA molecules is therefore very much reduced. Three phages in the Charon series that have been modified in this way are Charons 3A, 4A and 16A which differ from Charons 3, 4 and 16 by the amber mutations in genes *A* and *B*. The bacterial host strain, DP50supF, selected for use with the Charon phages, is a derivative of a strain χ1953 into which the *sup*F58 (suIII) suppressor mutation has been introduced. DP50supF is like the strain χ1776, developed with similar reasons in mind for plasmid cloning. The growth of both these strains is absolutely dependent upon the presence of diaminopimelic acid and thymidine in the culture medium. The metabolite diaminopimelic acid is not found in the mammalian gastrointestinal tract. The strains are also sensitive to bile salts and detergents and are quite difficult to grow even in the laboratory. The phages Charon 3A, 4A and 16A are confined to the lytic mode by deletions in the immunity regions, and by the *nin*5 deletion which results in the 'N-independent' activation of gene *Q* and thereby the genes of the late transcriptional pathway. Charon 4A and 16A recombinant molecules also lack *int* and *att* and so are unable to form lysogens.

Derivatives of the phage of Thomas *et al.* [16] have also been commonly used in experiments requiring biological containment [28]. These derivatives have amber mutations in genes *W*, *E* and *S*. They require the *sup*F suppressor in the bacterial host and so once again *E. coli* DP50supF is suitable. The phage λgtWES.λB has a deletion of the *Eco*RI fragment lying between sites 2 and 3, and so it is att^- and int^-, and cannot lysogenize. (The replaceable *Eco*RI fragment in these vectors contains the only cleavage sites for *Sst*I within the phage genome. The background of parental phage to recombinant phage is therefore greatly reduced if the vector DNA is cleaved with *Eco*RI and *Sst*I before carrying out the ligation reaction.)

(*c*) *Increased recovery of recombinants by* in vitro *packaging*

The transfection of phage DNA into *E. coli* cells rendered competent by treatment with $CaCl_2$ as described in Chapter 1 is an inefficient process giving between 10^5 and 10^6 plaques per μg of phage DNA (5×10^{-6}–5×10^{-5} plaques per DNA molecule). Hohn and Murray [29] and Sternberg *et al.* [30] have developed an *in vitro* packaging system which can yield 10^8 plaques per μg of DNA. In this system, the DNA to be packaged is mixed together with concentrated lysates from two cultures; one of cells which are undergoing infection with a phage whose maturation is blocked by an amber mutation in gene *D*, for example, and the other in which the infecting phage has a mutation in gene *E*, for example. The two lysates complement each other *in vitro*, and encapsidate concatemers of λ DNA to form mature phage. In practice the packaging extracts are made from *E. coli* strains lysogenized by a phage with the *cI*857 mutation. Cultures of the two lysogens are induced by temperature shift but cell lysis is blocked by a mutation in the *S* gene. The cells containing the unassembled head proteins can then be concentrated by centrifugation. Extracts are made by either sonicating such a cell pellet or subjecting it to a freeze–thaw cycle. Recombinant DNA can be mixed with the two extracts whereupon it is packaged to produce viable phage.

3.1.4 The construction and screening of libraries of recombinant DNAs in phage λ

It is easily possible to clone a set of DNA segments which are fully representative of all the DNA sequences within a complex genome. Such a collection of recombinant bacteriophage is known as a 'genomic library'. The number of clones necessary to represent all genomic sequences in such a library may be calculated from the following equation [29]

$$N = \ln(1-P)/\ln(1-f)$$

where N is the number of recombinants needed to have the probability, P, of isolating a gene which represents the fractional proportion of the genome, f. Thus for the average mammalian genome that has a haploid DNA content of about 3×10^6 kb, it is necessary to screen 8.1×10^5 phage in order to have a 99% probability ($P = 0.99$) of isolating a single copy gene contained in 17 kb cloned segments ($f = 1.7 \times 10^4 / 3 \times 10^9$).

The *in vitro* packaging systems described above are therefore a key requirement in the efficient construction and screening of libraries. These procedures are easily two orders of magnitude more efficient than the original procedures for transfection and therefore effectively mean that one can use proportionally less DNA to construct the library. Most of the replacement vectors have markers which permit recombinants to be distinguished from parental phage by plaque morphology. However, in order to ensure a high proportion of recombinants it is desirable to purify the left and right arms of the vector away from the replaceable segment. This may be done by allowing the vector DNA to anneal through its cohesive ends and then treating with DNA ligase to form covalently closed circles. Restriction endonuclease cleavage will then generate a fragment consisting of the joined left and right λ ends, which is easily separated from the replaceable segments by sucrose gradient sedimentation (Fig. 3.9). This approach was used by Maniatis *et al.* [32] in their construction of libraries of eukaryotic chromosomal segments (Chapter 2). These methods are now greatly simplified by the use of vectors, such as λEMBL4. This phage can be used as a vector for *Bam*HI fragments, and also for restriction fragments generated by enzymes such as *Sau*3A which have the same cohesive end. *Sau*3A has a tetranucleotide recognition sequence and so a partial digestion of chromosomal DNA with *Sau*3A approaches random cleavage more

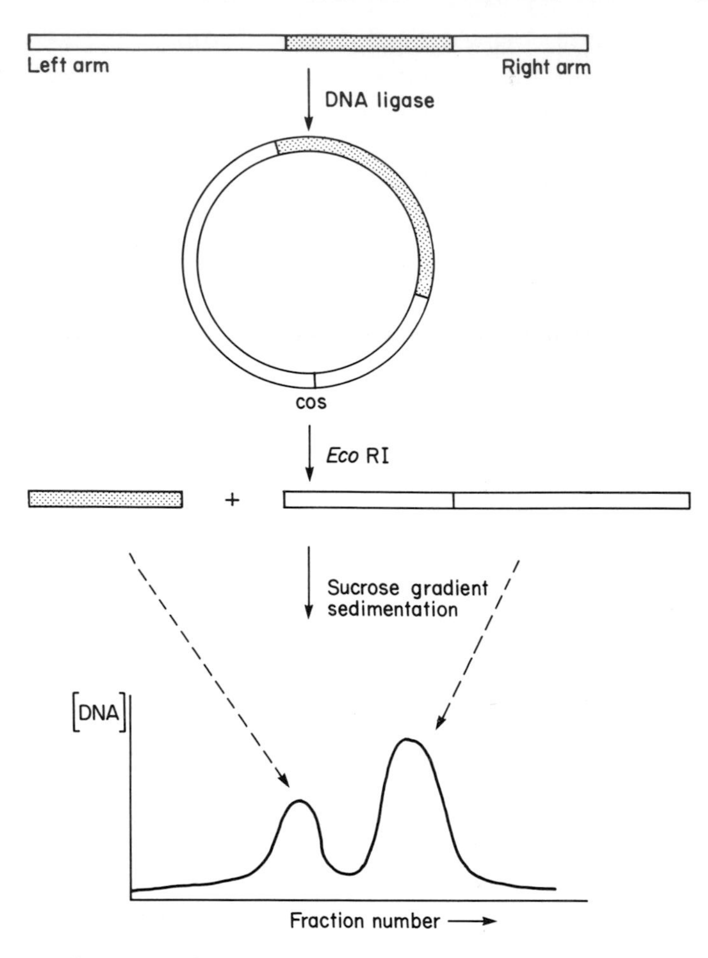

Figure 3.9 Purification of phage arms for cloning.

than would a partial digestion with an enzyme which recognizes a hexanucleotide sequence, for the reasons discussed in Chapter 2. Recombinants made from *Sau*3A fragments can be entirely removed from EMBL4 as large restriction fragments, with *Eco*RI which cleaves next to the *Bam*HI cloning site (see Fig. 3.7). This approach avoids the complex set of enzyme reactions which were used by Maniatis *et al.* [32]. Furthermore, the *Spi* selection offered by λEMBL4 removes the necessity for purifying the phage arms away from the central fragment before making the recombinant. Parental phage will be reconstituted in the ligation reaction but these will not grow on P2 lysogens (see Section 3.1.2(d)).

Large numbers of recombinant phage can be easily screened for complementarity towards a given radiolabelled sequence by the nucleic acid hybridization procedure of Benton and Davis [33]. The principle of the procedure is the same as that of colony hybridization (see Chapter 1). A nitrocellulose filter is laid onto the *E. coli* lawn and phage are absorbed onto the filter from the plaques. DNA is released from the phage on the filter and denatured by alkali treatment. The filter is brought to neutral pH, allowing single-stranded DNA to stick to the filter in the position of the plaque. The filter can then be incubated with a radiolabelled probe for a specific nucleotide sequence from the gene of interest. The position of hybridization of the probe is located by autoradiography and so the corresponding plaque can be picked from the master plate. This procedure permits the rapid screening of several hundred thousand plaques and so facilitates the isolation of single copy genes from complex eukaryotic genomes (see also Chapter 6).

An alternative screening method has been developed which permits the recovery of recombinant phage homologous to a sequence carried on a plasmid [34]. The method relies on homologous recombination occurring *in vivo* between the phage, which could be a member of a genome library and the plasmid, which could for example, contain a cloned cDNA (Fig. 3.10). The plasmid, πVX, is 902 base pairs in length and consists of the ColEl replication origin, a polylinker with cloning sites for the fragments generated by seven enzymes, and the suppressor tRNA gene *sup*F. The bacterial host chosen for propagating this plasmid contains a second plasmid carrying ampicillin and tetracycline genes with amber mutations. The plasmid, or its recombinants, can be introduced into this strain by transfection and antibiotic selection applied to identify those cells in which the amber mutations are suppressed by the incoming plasmid. The phage library is constructed in a phage vector with amber mutations in essential genes, and is propagated on cells carrying the plasmid. Recombination between homologous sequences cloned in the plasmid and phage generates phages which carry an integrated copy of the plasmid. Such recombinants can be selectively purified by their ability to grow in suppressor free hosts, unlike the remainder of the phage library. The one problem of this approach is that the cloned segment of DNA selected from the phage library contains a duplication of the sequence carried on the plasmid together with the inserted plasmid. The plasmid can excise by the reciprocal homologous recombination event, but there is no efficient selection for this. Furthermore, the technique should be applied with caution for complex eukaryotic genomes, in which many so-called 'unique' genes are members of small families of closely related genes, which sometimes occur in clusters and are sometimes dispersed over the genome. The technique may then be put to good use to isolate members of such a gene family but, at the same time, the possibility of rearrangement of the genomic DNA by homologous recombination

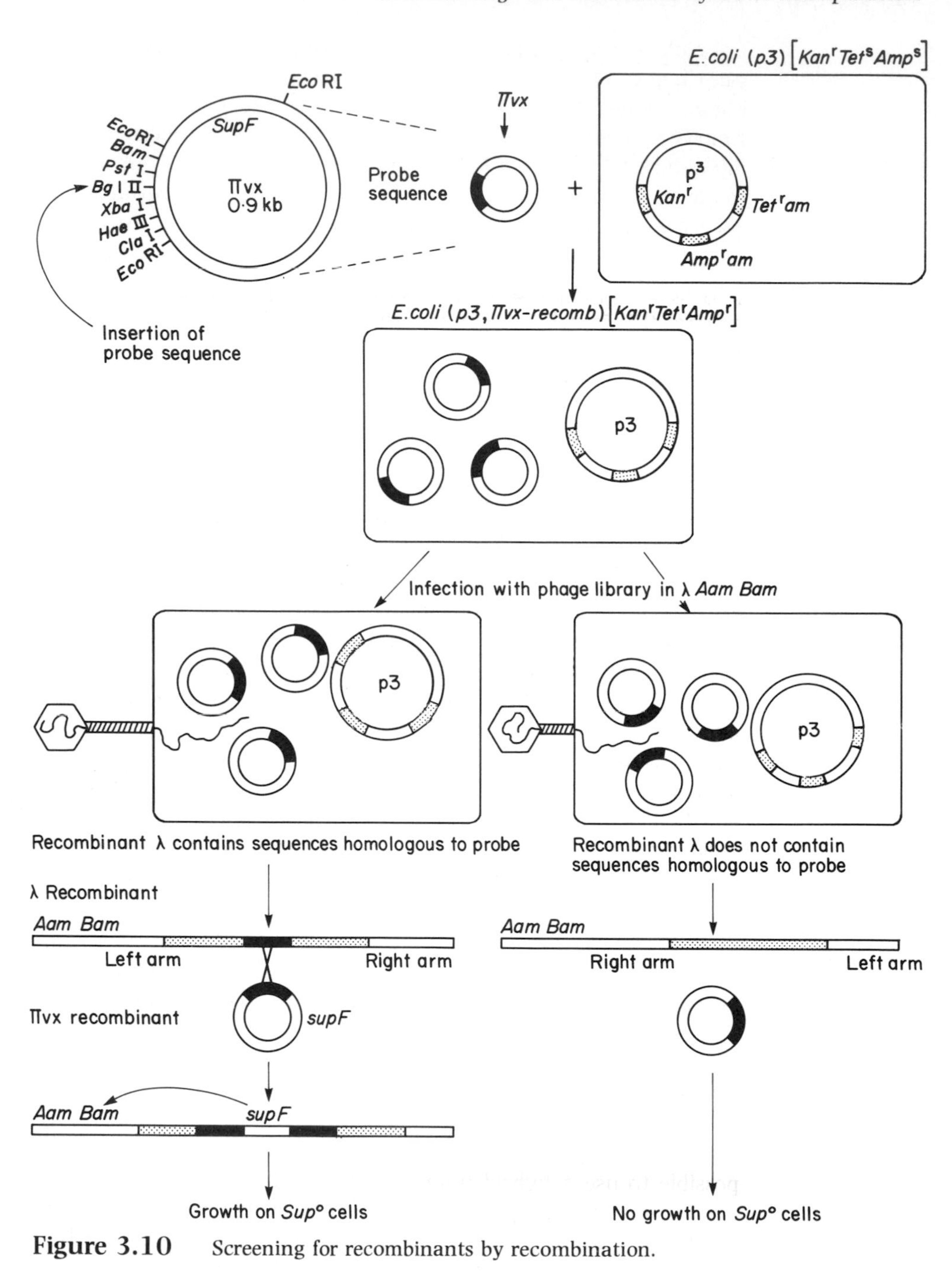

Figure 3.10 Screening for recombinants by recombination.

should be a note of caution to be considered with respect to any given experiment.

3.1.5 Expression from phage λ promoters

Phage λ has been successfully used as a means of amplifying gene copy number in order to maximize expression from the promoter of that gene. This has been done for the genes for DNA ligase [35], DNA polymerase I [36] and the γ subunit of DNA polymerase III [37]. In general this is achieved by incorporating a Q^- mutation into the phage genome in order to block the expression of late genes and so prolong DNA replication. For convenience, these recombinant phages usually carry a thermosensitive mutation in the *cI* gene. This enables the phage genome to be maintained as a prophage. The lysogenic strain can then be conveniently induced by shifting the temperature of a large culture. Most phage vectors accept DNA in a region of the genome where either the p_L promoter or the p'_R promoter could be used for efficient transcription. The work of Murray *et al.* [38] is an ideal case study of the comparative yields of T4 DNA ligase from a series of λ-T4 recombinants constructed *in vitro* (T4*lig* phages) in which transcription is either from T4 promoters or from the p_L and p_R promoters. The genomes of the λ-T4*lig* phages used in this study are shown in Fig. 3.11. Expression of the T4 ligase gene in phages λNM873, λNM875 and λNM993 occurs from T4 promoters when the major λ promoters are repressed. The amplification of the enzyme yield depends upon the gene dosage, which increases during the lytic interaction with the host. The lytic interaction can be initiated by thermoinduction of lysogens since the *cI* gene contains a temperature-sensitive mutation. The incorporation of mutations in *Q* and *S* drastically reduces late transcription and prevents cell lysis and so allows a high yield of phage DNA and retention of all phage-determined polypeptides within the cell. The production of ligase from such phage is not, however, particularly efficient.

The phage λNM1001 and λNM967 are phage in which the ligase gene should be transcribed from p_L. Ideally the phage should be cro^- to relieve the repressive effect of the *cro* gene product on transcription from p_L and p_R. Since cro^- phage replicate their DNA poorly, it is possible to use a hybrid immunity phage that contains p_R^{434} and cro^{434} along with the p_L of λ which is not repressed by the cro^{434} gene product. Such a hybrid immunity phage can only undergo lytic interactions with the host. The most efficient amplification of T4 ligase was obtained from the induction of the λNM989 prophage in which transcription is from p'_R. Efficient late transcription from this promoter requires the *Q* gene product and so in this case the *Q* is wild

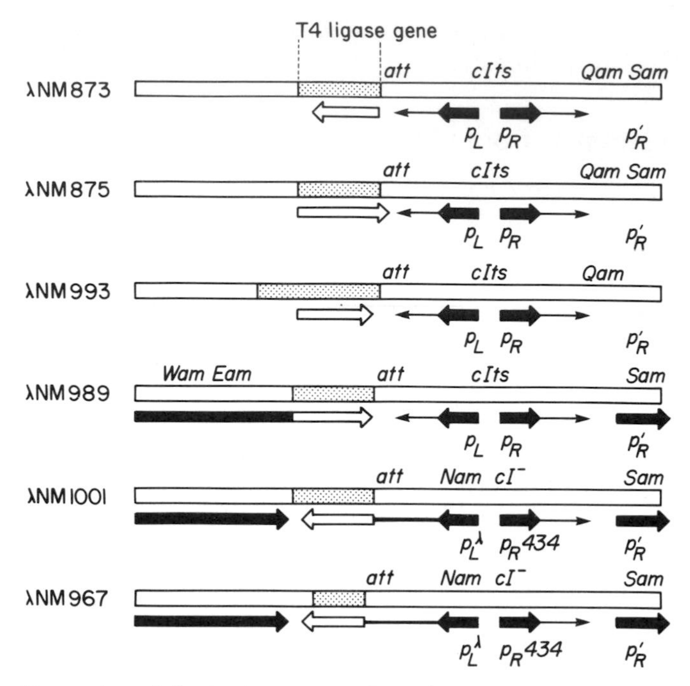

Figure 3.11 Expression of the T4 ligase gene from λ promoters.

type. Cell lysis and packaging are eliminated by mutations in genes *S*, *W* and *E* (see Section 3.1.3).

The p_L promoter of bacteriophage has also been incorporated into several plasmid expression vectors [39, 40]. The promoter in these plasmids is under the control of thermo-sensitive repressor which can be encoded either by the plasmid or a prophage in the host cell.

3.1.6 Cosmids

Cosmids were first developed by Collins and Hohn [41]. As the name suggests they are hybrid vectors comprised in part of plasmid sequences but containing the *cos* site of phage λ. The plasmid sequences consist of a replication origin and one or two selectable markers containing unique restriction sites for cloning. Cosmids allow the cloning of 40 kb segments of foreign DNA by the principles shown in Fig. 3.12. High molecular weight segments of the foreign DNA, generated by restriction endonuclease cleavage, are incubated with the linear cosmid molecule at high DNA concentration to favour the production of oligomers. A complex mixture of ligation products will be produced (as shown in Fig. 3.12) but amongst these will be molecules in which the 40 kb segment of foreign DNA is flanked by

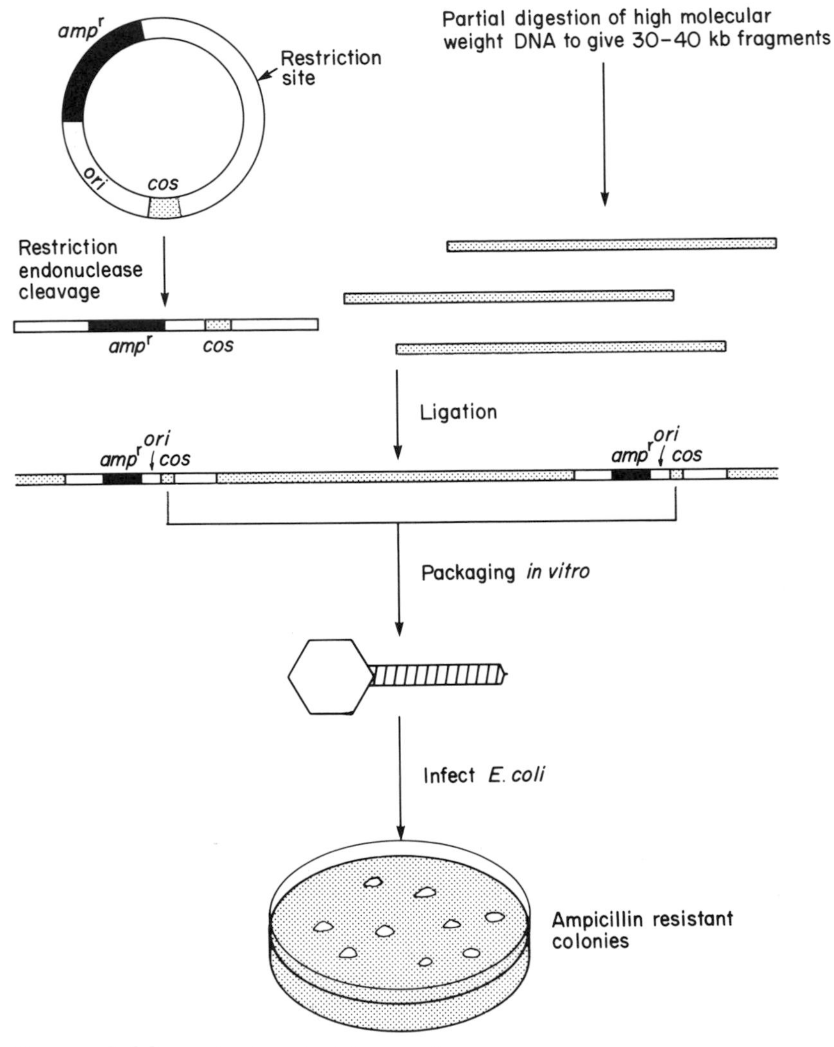

Figure 3.12 Cloning DNA in cosmids.

two cosmid molecules and hence *cos* sites in the same orientation. These structures resemble the molecules produced late in phage infections and are a suitable substrate for the *ter* function of the phage A protein. The foreign DNA which intervenes between the *cos* sites can therefore be packaged into bacteriophage particles in an *in vitro* packaging reaction, whereupon the structure is cleaved to give a linear molecule with cohesive ends. Upon infection of *E. coli*, the injected recombinant cyclizes through the *cos* sites and then replicates as a plasmid and expresses the drug resistance marker. A number of cosmid vectors have been developed which are based upon the commonly used plasmid vectors [42, 43]. Subsequently, selectable eukaryotic *markers* (see Chapter 8) have been introduced into the cosmids so thay can act as shuttle vectors which permit the propagation of the cloned DNA either in *E. coli* or in mammalian cells [44].

3.2 Single-stranded DNA phages

3.2.1 The life cycle of Ml3

The filamentous coliphages Ml3, fd and f1 are a set of closely related bacteriophages which contain highly homologous single-stranded circular DNA molecules 6.4 kb long. The infection cycle of Ml3 is illustrated in Fig. 3.13. The DNA is ensheathed in a tube of some

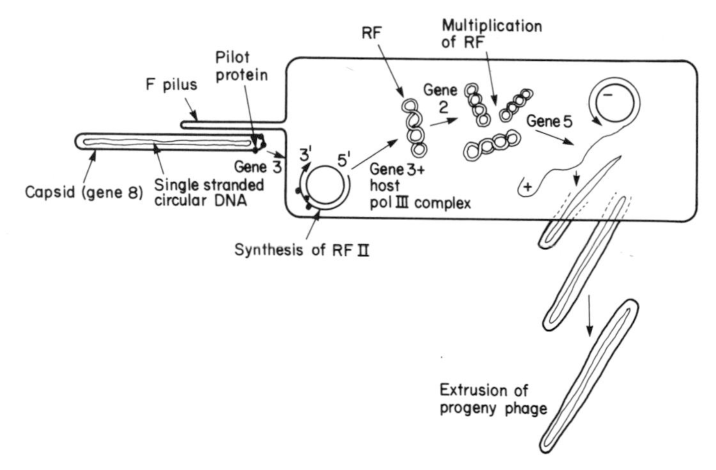

Figure 3.13 The infection cycle of Ml3.

3000 molecules of the major capsid protein, encoded by the phage gene 8. The phage is specific for male bacteria and binds to receptors on the sex pili specified by the F plasmid. Phage DNA can, however, be introduced into female cells by transfection. The incoming phage DNA is converted to a double-stranded replicative form (RF) upon penetration of the virus. This process requires the product of gene 3, known as the pilot protein since it is present in the capsid and 'pilots' the phage DNA into the cell. A detailed description of the replication of the phage DNA may be found elsewhere [45]. Synthesis of the RF from the single-stranded viral DNA is initiated from an RNA primer and extended by the DNA polymerase III* system of the host cell. The subsequent replication of the RF to give 50–100 RF molecules per cell requires the product of gene 2. The pattern of replication changes at this point. Phage encoded single-stranded binding protein (the product of gene 5) binds to the viral strands which are synthesized by rolling circle replication. The progeny molecules are packaged upon extrusion through the cell wall. The cells are not killed but grow much more slowly than uninfected cells. The 'plaques' are therefore highly turbid in appearance. Phage particles can be recovered in high yield from the culture supernatant by precipitation with polyethylene glycol. There is no defined upper limit to the size of DNA that can be packaged into Ml3 particles, but in practice it is found that recombinant phage carrying more than 5 kb of foreign DNA are unstable. The double-stranded RF can be prepared from infected bacteria and handled exactly like a plasmid in terms of constructing recombinant DNA molecules. Single-stranded DNA can be prepared by the deproteinization of virus preparations.

3.2.2 The development of Ml3 vectors

In order to use Ml3 as a cloning vector it was first necessary to identify a non-essential region into which one could insert foreign DNA. The complete sequence of Ml3 has been determined and only one intergenic region has been recognized. This region is 507 nucleotides long and is located between genes 2 and 4. The region contains the origin of replication, but it was categorically shown that the integrity of the whole region is not necessary for phage development by Messing and his colleagues who introduced a segment of the *E. coli lac* operon into this site (Fig. 3.14). This was the first stage in the development of a series of Ml3 cloning vectors [46, 47, 48]. The phage DNA has ten cleavage sites for *Bsu*I, one of which is located in the intergenic region. A *Hin*II fragment containing the regulatory region of the *E. coli lac* operon and the first 145 codons of the β-galactosidase gene was inserted into this site. The phage RF DNA was partially digested with *Bsu*I and full length linear monomers were ligated to the *lac Hin*II fragment. Only those recombinants in which the *lac* DNA is inserted into a non-essential region are viable. The segment of the *lac* gene contained on this fragment encodes the α peptide, which although enzymically inactive will complement the amino-terminally deleted β-galactosidase specified by the *lac*Z *15* deletion in the host *E. coli*. Functional β-galactosidase can then be detected by the ability of cells infected by the recombinant phage to generate blue plaques on plates containing the inducer IPTG and the chromogenic substrate X-gal. Subsequently, a series of cloning vectors have been developed in which the insertion of foreign DNA inactivates the gene for the α peptide. This enables phage recombinants to be identified from the white plaques they produce on bacterial lawns plated on media containing X-gal.

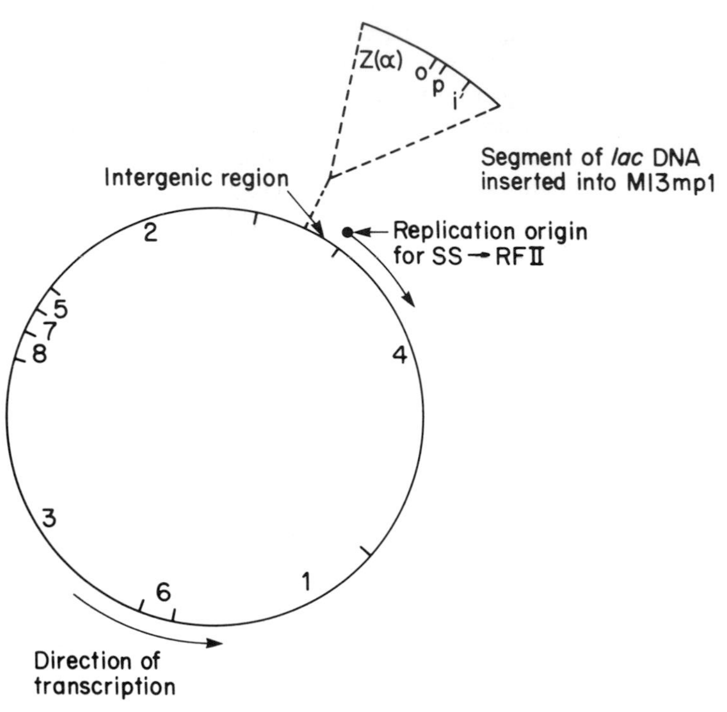

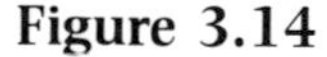
Figure 3.14 Map of the Ml3 DNA.

Ml3mp1 does not contain any unique restriction sites which are particularly useful for gene cloning. It was known from nucleotide sequence data that the sequence GGATTC occurred at a site corresponding to the fifth amino-acid of β-galactosidase. Gronenborn and Messing [46] reasoned that a single mutation changing G to A would create an *Eco*RI site in the peptide gene without affecting its biological function. This was achieved by methylating guanine residues in the DNA with methyl-*N*-nitrosourea. The methylated Gs will then mispair with Ts during replication. Sufficient alkylating agent was used to methylate two to four guanine residues per genome and the DNA was then transfected into bacterial cells. After several cycles of infection, phage RF DNA was recovered and cleaved with *Eco*RI. Linear molecules were prepared by gel electrophoresis, religated and retransfected into *E. coli*. Two mutant phage were isolated in this way, Ml3mp2 and Ml3mp3. These have *Eco*RI sites at the position corresponding to the β-galactosidase amino-acids 5 and 119 respectively.

The restriction site repertoire of Ml3mp2 has subsequently been extended by the addition of a set of chemically synthesized linkers into this *Eco*RI site. Initially an oligonucleotide was inserted which had cleavage sites for the enzymes *Bam*HI, *Sal*I, *Acc*I, *Hinc*II, *Pst*I, and was flanked by *Eco*RI sites. This sequence contributes 14 extra codons to the β-galactosidase gene but does not affect the ability of the peptide to undergo intracistronic complementation. There are, however, additional sites for *Bam*HI in gene 3 and for *Hinc*II and *Acc*I in gene 2. These sites were removed in a manner analogous to that described above for the introduction of the *Eco*RI site. The phage DNA was mutagenized, propagated in *E. coli* and the mutant phage RF DNA selected from molecules resistant to the appropriate restriction endonuclease. The mutagenesis was planned in such a way that the nucleotide change would be in the third base positions of the codons concerned and so would not change the amino-acid sequence of the gene products. The *Bam* site was mutated by methyl-*N*-nitrosourea treatment and the *Hinc*II and *Acc*I sites by mutagenesis with hydroxylamine. These mutations were then transferred onto the phage genome carrying the synthetic 17-mer cloning site. This was achieved by isolating a restriction fragment containing the mutated phage genes 2 and 3 from RF DNA. This fragment was denatured and annealed to the + viral strand of the other phage DNA. The partial heteroduplexes formed in this way replicate following their introduction into *E. coli* [47]. The resulting vector M13mp7 is still infectious and still directs the synthesis of functional α-peptide, unless the reading frame is disrupted by the insertion of DNA into the oligonucleotide. The geneology of this series of M13 vectors is shown in Fig. 3.15.

This vector has now been developed to include cloning sites for *Hind*III, *Sma*I and *Xma*I [48]. Two phage have been developed (M13mp8 and M13mp9) in which the modified multiple restriction site region is arranged in opposite orientations with respect to the M13 genome (Fig. 3.16). This means that any given restriction fragment bounded by a pair of these sites can be cloned in both orientations. This is particularly useful for DNA sequencing applications since it enables a nucleotide sequence to be determined in both directions, thereby generating overlapping sequence data.

A series of plasmids (the pEMBL plasmids) have now been constructed which can be propagated as double-stranded circular DNA molecules, or alternatively packaged as single-stranded circular DNA into the capsids of filamentous phage [49]. The plasmids contain the β-lactamase gene as a selectable marker (see Chapter 4); the segment of the *lac*Z gene encoding the α-peptide; and a segment of the f1 genome containing all the cis-acting elements needed for DNA replication and morphogenesis. The pEMBL plasmids are handled, for cloning purposes, exactly as other plasmids. A cell containing one of these plasmids can, however, be super-infected with f1 phage, whereupon both the pEMBL genome and the super-infecting phage will be replicated and packaged as single-stranded DNA.

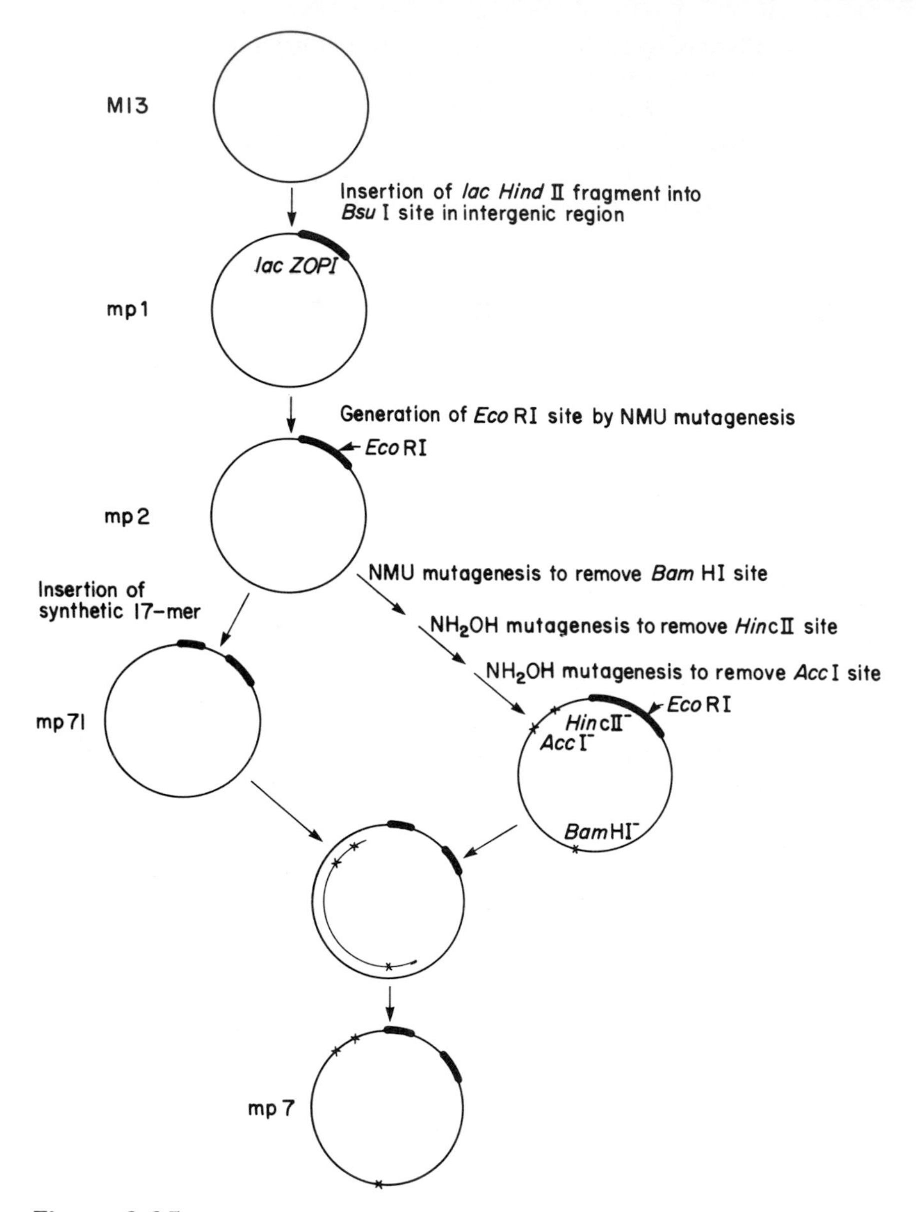

Figure 3.15 The geneology of Ml3 vectors.

3.2.3 Sequencing DNA cloned in Ml3

DNA cloned in Ml3 can be readily sequenced using the chain terminator procedure developed by Sanger and his colleagues [50]. Oligonucleotide primers are available which are complementary to the sequence immediately adjacent to the cloning sites of the Messing

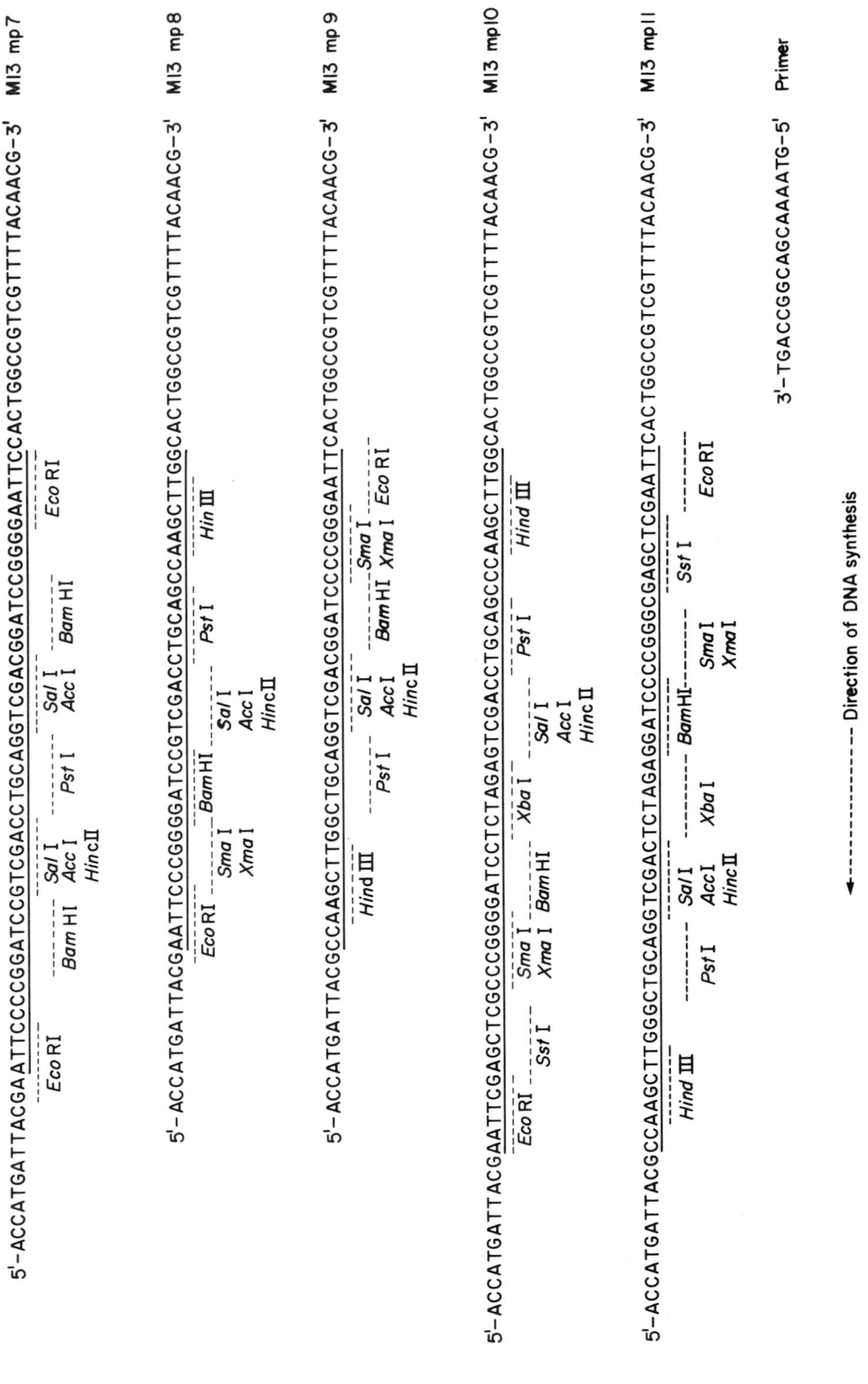

5'-ACCATGATTACGAATTCCCCGGATCCGTCGACCTGCAGGTCGACGGATCCGGGGAATTCCACTGGCCGTCGTTTTACAACG-3' M13 mp7

Eco RI, *Bam* HI, *Sal* I / *Acc* I / *Hinc* II, *Pst* I, *Sal* I / *Acc* I, *Bam* HI, *Eco* RI

5'-ACCATGATTACGAATTCCCGGGGATCCGTCGACCTGCAGCCAAGCTTGGCACTGGCCGTCGTTTTACAACG-3' M13 mp8

Eco RI, *Sma* I / *Xma* I, *Bam* HI, *Sal* I / *Acc* I / *Hinc* II, *Pst* I, *Hin* III

5'-ACCATGATTACGCCAAGCTTGGCTGCAGGTCGACGGATCCCCGGGAATTCACTGGCCGTCGTTTTACAACG-3' M13 mp9

Hind III, *Pst* I, *Sal* I / *Acc* I / *Hinc* II, *Bam* HI, *Sma* I / *Xma* I, *Eco* RI

5'-ACCATGATTACGAATTCGAGCTCGCCCGGGGATCCTCTAGAGTCGACCTGCAGCCCAAGCTTGGCACTGGCCGTCGTTTTACAACG-3' M13 mp10

Eco RI, *Sst* I, *Sma* I / *Xma* I, *Bam* HI, *Xba* I, *Sal* I / *Acc* I / *Hinc* II, *Pst* I, *Hind* III

5'-ACCATGATTACGCCAAGCTTGGGCTGCAGGTCGACTCTAGAGGATCCCCGGGCGAGCTCGAATTCACTGGCCGTCGTTTTACAACG-3' M13 mp11

Hind III, *Pst* I, *Sal* I / *Acc* I / *Hinc* II, *Xba* I, *Bam* HI, *Sma* I / *Xma* I, *Sst* I, *Eco* RI

3'-TGACCGGCAGCAAAATG-5' Primer

←---------------- Direction of DNA synthesis

Figure 3.16 Multiple restriction sites in M13 vectors.

vectors. The primer is annealed to the single-stranded recombinant DNA and four DNA synthesis reactions are carried out using the Klenow fragment of *E. coli* DNA polymerase I. These reactions each contain the four deoxynucleoside triphosphates one of which is labelled with ^{32}P (or more recently a ^{35}S-thio analogue has been used). Each of the four reactions contains a low concentration of the 2′ 3′ dideoxy analogue of one of the deoxynucleoside triphosphates. When this analogue is incorporated into DNA, the chain can no longer be extended since it lacks a 3′OH group. In the reaction mixture containing dideoxy-GTP, for example, a proportion of the chain elongation reactions will terminate each time a C residue is encountered in the template. The product of the reaction will therefore be a population of ^{32}P labelled molecules having a common 5′ end, but varying in length to each successive 3′ G residue in the sequence. When fractionated by polyacrylamide gel electrophoresis, these fragments can be visualized autoradiographically. The gel systems used can fractionate oligonucleotides differing in length by one nucleotide up to several hundred residues long. By running the products of the four reaction mixtures alongside each other, it is therefore possible to determine the successive order of nucleotides 3′ to the common 5′ terminus (Fig. 3.17).

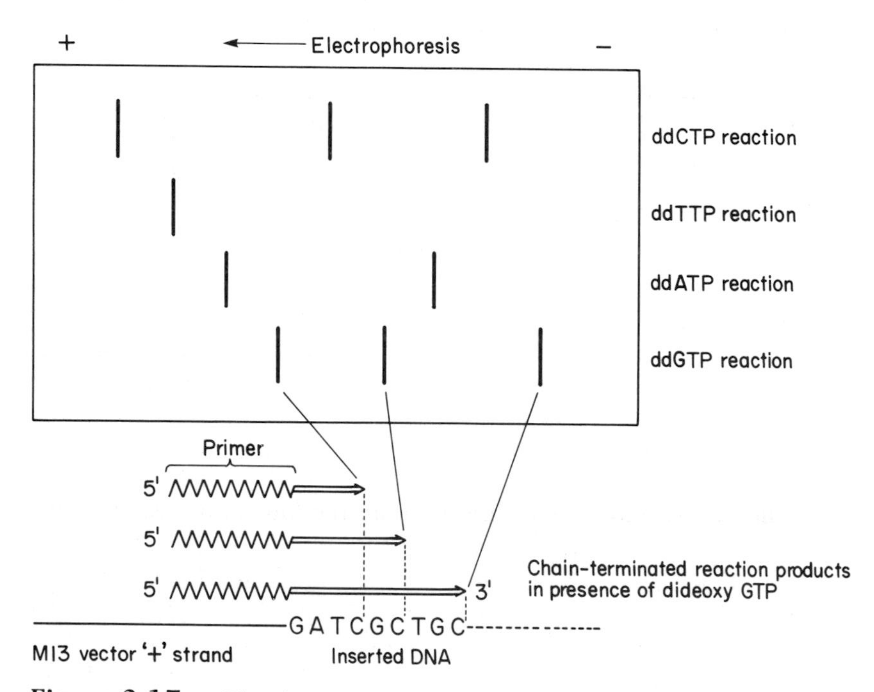

Figure 3.17 The chain-terminator DNA sequencing technique.

3.2.4 Mutagenesis of DNA cloned in single-stranded DNA phage vectors

We have already encountered methods for introducing mutations into M13 genomes in the discussion on the development of M13 cloning vectors. Sodium bisulphite, which deaminates cytosine bases in single-stranded DNA to give uracil, has also been used for the general mutagenesis of the entire recombinant genome. One approach has used such mutagenized viral DNA as a template for DNA synthesis. Restriction fragments have then been cleaved from the resulting double-stranded DNA and recloned [51]. An alternative approach has been to expose single-stranded DNA in a heteroduplex structure formed by taking denatured DNA from the RF of a deletion mutant cleaved at a unique restriction site and annealing it with the equivalent molecule containing the corresponding wild type region. This heteroduplex structure was subjected to bisulphite mutagenesis and then introduced into *E. coli*. This procedure results in a high proportion of phage with point mutations in the region corresponding to the deletion mutant [52].

An extremely powerful method of site directed mutagenesis utilizes a chemically synthesized oligonucleotide containing the mutation as primer for elongation by the Klenow fragment of *E. coli* DNA polymerase [53, 54]. In the presence of T4 DNA ligase and ATP, covalently closed circles are formed on the completion of DNA synthesis. These molecules can then be introduced into *E. coli*. The method has been used to produce both point mutations and deletions. In practice oligomers of 10–12 nucleotides have been used to produce stable primers containing a single mismatch. Longer oligomers are needed to stabilize deletion loops formed by the primer. Wallace *et al.* [53] used a 21 nucleotide long primer, for example, to make a 14 nucleotide deletion in a yeast tRNA gene. Recovery of mutants can be low as a result of excision repair of the mismatch. The recombinants can, however, be screened by plaque hybridization with the labelled oligonucleotide used as primer (see Fig. 3.18). The mutant plaques will hybridize more efficiently to this probe than will wild type plaques. The two types of plaque may therefore be distinguished since hybrids between the wild type plaque and the labelled oligonucleotide will melt at a lower temperature than the fully annealed structure formed by DNA in the mutant plaques and the oligonucleotide. Alternatively, progeny phage can be isolated and its DNA used in another reaction with the same primer. The mutated template will give more efficient primed synthesis than the wild type template, so the mutant phage will be enriched [54]. The efficiency of generating mutants can also be increased by methylating the strand containing

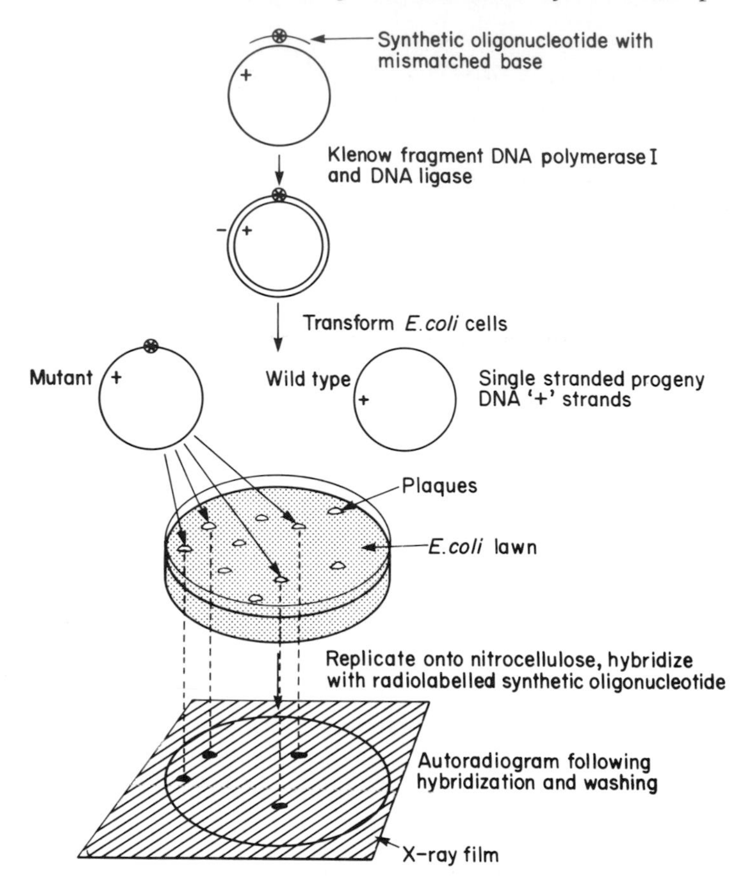

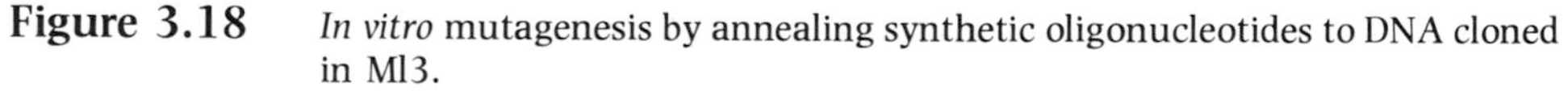

Figure 3.18 *In vitro* mutagenesis by annealing synthetic oligonucleotides to DNA cloned in Ml3.

the mutation in the *N*6 adenine position within a GATC sequence [55] and by using non-methylated phage DNA (grown in adenine methylase deficient, *dam*$^-$ *E. coli*) as template. The rationale for this approach is that mismatch repair *in vivo* shows a preference for the unmethylated DNA strand [56].

One of the ultimate aims of genetic engineering must be the application of the techniques of *in vitro* mutagenesis to the design of novel enzymes. This line of work is in its infancy, but the first steps have been taken with a study of directed mutations which change amino-acids at the ATP binding site of the tyrosyl-tRNA synthetase from *Bacillus stearothermophilus* [57]. This group have used synthetic oligonucleotides to direct mutations in the gene for this enzyme cloned in Ml3. In this way they have made a set of mutants in which those residues which undergo hydrogen bonding with the substrate have been systematically altered. Knowledge of the three dimensional structure of this protein from X-ray crystallographic data makes this an extremely powerful approach to redesigning enzymes. The potential which this technique has for improving the properties of enzymes of industrial importance remains to be realized.

References

1. Hershey, A. D. (ed.) (1971) *The Bacteriophage Lambda*. Cold Spring Harbor Laboratory, New York.
2. Lewin, B. (1977) *Gene Expression III*, John Wiley, New York.
3. Wu, R. and Taylor, E. (1971) Nucleotide sequence analysis of DNA: complete nucleotide sequence of the cohesive ends of bacteriophage λ DNA. *J. Mol. Biol.*, **57**, 491–511.
4. Inman, R. B. (1966) A denaturation map of the λ phage DNA molecule determined by electron microscopy. *J. Mol. Biol.*, **18**, 464–76.
5. Bastia, D., Sneoka, N. and Cos, E. C. (1975) Studies on the late replication of phage lambda: rolling circle replication of the wild type and a partially suppressed strain, Oam29 Pam80. *J. Mol. Biol.*, **98**, 305–20.
6. Enquist, L. W. and Skalka, A. (1973) Replication of bacteriophage λ DNA dependent on the function of host and viral genes. Interaction of *red*, *gam* and *rec*. *J. Mol. Biol.*, **75**, 185–212.
7. Wang, J. C. and Kaiser, A. D. (1973) Evidence that the cohesive ends of mature λ DNA are generated by the gene A product. *Nature New Biol.*, **241**, 16–17.
8. Roberts, J. (1969) Termination factor for RNA synthesis. *Nature*, **224**, 1168–74.
9. Adhya, S., Gottesman, M. and De Crombrugghe, B. (1974) Release of polarity in *Escherichia coli* by gene N of phage λ, termination and antitermination of transcription. *Proc. Natn Acad. Sci. USA*, **71**, 2534–8.
10. Franklin, N. (1974) Altered reading of genetic signals fused to the *N* operon of bacteriophage λ: genetic evidence for modification of polymerase by the protein product of the *N* gene. *J. Mol. Biol.*, **89**, 33–48.
11. Ptashne, M., Backman, K., Humaynn, M. Z. *et al.* (1976) Autoregulation and function of a repressor in bacteriophage λ. *Science*, **194**, 156–61.
12. Roberts, J. W., Roberts, C. W. and Mount, D. W. (1977) Inactivation and proteolytic cleavage of phage λ repressor *in vitro* in an ATP dependent reaction. *Proc. Natn Acad. Sci., USA*, **74**, 2283–7.
13. Simon, M. N., Davis, R. W. and Davidson, N. (1971) In *The Bacteriophage Lambda* (ed. A. D. Hershey), Cold Spring Harbor Laboratory, New York.
14. Rambach, A. and Tiollais, P. (1974) Bacteriophage λ having *Eco*RI nuclease sites only in the non-essential region of the genome. *Proc. Natn. Acad. Sci. USA*, **71**, 3927–30.
15. Murray, N. E. and Murray, K. (1975) Manipulation of restriction targets in phage λ to form receptor chromosomes for DNA fragments. *Nature*, **251**, 476–81.
16. Thomas, M., Cameron, J. R. and Davis, R. W. (1974) Viable molecular hybrids of bacteriophage λ and eukaryotic DNA. *Proc. Natn Acad. Sci. USA*, **71**, 4579–83.

17. Murray, N. E., Brammar, W. J. and Murray, K. (1977) Lambdoid phages that simplify the recovery of *in vitro* recombinants. *Mol. Gen. Genet.*, **150**, 53–61.
18. Blattner, F. R., Williams, B. G., Blechl, A. E. *et al.* (1977) Charon phages: safer derivatives of bacteriophage λ for DNA cloning. *Science*, **196**, 161–9.
19. Lathe, R. and Lecocq, J.-P. (1977) Overproduction of a viral protein during infection of a *lyc* mutant of *Escherichia coli* with phage λ *imm*434. *Virology*, **83**, 204–6.
20. Scherer, G., Telford, J., Baldari, C. and Pirrotta, V. (1981) Isolation of cloned genes differentially expressed at early and late stages of *Drosophila* embryonic development. *Devl Biol.*, **86**, 438–47.
21. Young, R. A. and Davis, R. W. (1983) Efficient isolation of genes by using antibody probes. *Proc. Natn Acad. Sci. USA*, **80**, 1194–8.
22. Kemp, D. J., Coppell, R. L., Cowman, A. F. *et al.* (1983) Expression of *Plasmodium falciparum* blood stage antigens in *E. coli*: detection with antibodies from immune humans. *Proc. Natn Acad. Sci. USA*, **80**, 3787–91.
23. Loenen, W. A. M. and Brammar, W. J. (1980) A bacteriophage λ vector for cloning large DNA segments made with several restriction enzymes. *Gene*, **10**, 249–57.
24. Karn, J., Brenner, S., Barnett, L. and Cesareni, G. (1980) A novel bacteriophage λ cloning vector. *Proc. Natn Acad. Sci USA*, **77**, 5172–6.
25. Murray, N. E. (1983) in *The Bacteriophage Lambda*, Vol. II, Cold Spring Harbor, Laboratory, New York.
26. Hohn, T., Wurtz, M. and Hohn, B. (1976) Capsid transformation during packaging of bacteriophage λ DNA. *Phil. Trans. R. Soc.*, **276**, 51–61.
27. Casjens, S. and King, J. (1975) Virus assembly. *A. Rev. Biochem.*, **44**, 555–611.
28. Leder, P., Tiemeier, D. and Enquist, L. (1977) EK2 derivatives of bacteriophage λ useful in the cloning of DNA from higher organisms: the λ gt.*WES* system. *Science*, **196**, 175–7.
29. Hohn, B. and Murray, K. (1977) Packaging recombinant DNA molecules into bacteriophage particles *in vitro*. *Proc. Natn Acad. Sci. USA*, **74**, 3259–63.
30. Sternberg, N., Teimeier, D. and Enquist, L. (1977) *In vitro* packaging of a λ-*dam* vector containing *Eco*RI DNA fragments of *Escherichia coli* and phage P1. *Gene*, **1**, 255–80.
31. Clarke, L. and Carbon, J. (1976) A colony bank containing synthetic ColEl hybrid plasmids representative of the entire *E. coli* genome. *Cell*, **9**, 91–9.
32. Maniatis, T., Hardison, R. C., Lacy, E. *et al.* The isolation of structural genes from libraries of eukaryotic DNA. *Cell*, **15**, 687–701.
33. Benton, W. and Davis, R. W. (1977) Screening λ-gt recombinant clones by hybridisation to single plaques *in situ*. *Science*, **196**, 180–2.
34. Seed, B. (1983) Purification of genomic sequences from bacteriophage libraries by recombination and selection *in vivo*. *Nucl. Acids Res.*, **11**, 2427–45.
35. Panasenko, S. N., Cameron, J. R., Davis, R. W. and Lehman, I. R. (1977)

500 fold overproduction of DNA ligase after induction of a hybrid λ lysogen constructed *in vitro*. *Science*, **196**, 188–9.
36. Murray, N. E. and Kelley, W. S. (1979) Characterisation of λ *polA* transducing phages; effective expression of the *E. coli polA* gene. *Mol. Gen. Genet.*, **175**, 77–87.
37. Hubscher, U. and Kornberg, A. (1980) The *dnaZ* protein, the γ-subunit of DNA polymerase 3 holoenzyme of *Escherichia coli*. *J. Biol. Chem.*, **255**, 11698–703.
38. Murray, N. E., Bruce, S. and Murray, K. (1979) Molecular cloning of the DNA ligase gene from bacteriophage T4. Amplification and preparation of the gene product. *J. Mol. Biol.*, **132**, 493–505.
39. Remaut, E., Stannsens, P. and Fiers, W. (1981) Plasmid vectors for high efficiency expression controlled by the p_L promoter of coliphage λ. *Gene*, **15**, 81–93.
40. Shimatake, H. and Rosenberg, M. (1981) Purified λ regulatory protein *c*II positively activates promoters for lysogenic development. *Nature*, **292**, 128–32.
41. Collins, J. and Hohn, B. (1978) Cosmids: a type of plasmid gene cloning vector that is packageable *in vitro* in bacteriophage λ heads. *Proc. Natn Acad. Sci. USA*, **75**, 4242–6.
42. Ish-Horowicz, D. and Burke, J. F. (1981) Rapid and efficient cosmid cloning. *Nucl. Acids Res.*, **9**, 2989–98.
43. Chia, W., Scott, M. R. D. and Rigby, P. W. J. (1982) The construction of cosmid libraries of eukaryotic DNA using the Homer series of vectors. *Nucl. Acids Res.*, **10**, 2503–20.
44. Grosveld, F. G., Land, T., Murray, E. J. *et al.* The construction of cosmid libraries which can be used to transform eukaryotic cells. *Nucl. Acids Res.*, **10**, 6715–32.
45. Kornberg, A. (1980) *DNA Replication*, W. H. Freeman, San Francisco.
46. Gronenborn, B. and Messing, J. (1978) Methylation of single-stranded DNA *in vitro* introduces new restriction endonuclease cleavage sites. *Nature*, **277**, 375–7.
47. Messing, J., Crea, R. and Seeberg, P. H. (1981) A system for shotgun DNA sequencing. *Nucl. Acids Res.*, **9**, 309–21.
48. Messing, J. and Vieira, J. (1982) A new pair of Ml3 vectors for selecting either DNA strands or double digest restriction fragments. *Gene*, **19**, 269–76.
49. Dente, L., Cesareni, G. and Cortese, R. (1983) pEMBL: a new family of single-stranded plasmids. *Nucl. Acids Res.*, **11**, 1645–55.
50. Sanger, F., Nicklen, S. and Coulson, A. R. (1979) DNA sequencing with chain terminating inhibitors. *Proc. Natn Acad. Sci. USA*, **74**, 5463–7.
51. Weiher, H. and Schaller, H. (1982) Segment specific mutagenesis: extensive mutagenesis of a *lac* promoter/operator element. *Proc. Natn Acad. Sci. USA*, **79**, 1408–12.
52. Everett, R. D. and Chambon, P. (1982) A rapid and efficient method for region specific and strand specific mutagenesis of cloned DNA. *Eur. Mol. Biol. Org.*, **1**, 433–7.
53. Wallace, R. B., Johnson, P. F., Tanaka, S. *et al.* (1980) Directed deletion of a yeast transfer RNA intervening sequence. *Science*, **209**, 1396–400.

54. Gillam, S., Astell, C. R. and Smith, M. (1980) Site specific mutagenesis using oligodeoxyribonucleotides: isolation of a phenotypically silent ϕX174 mutant, with a specific nucleotide deletion, at very high efficiency. *Gene*, **12**, 129–37.
55. Kramer, W., Schughart, K. and Fritz, H. J. (1982) Directed mutagenesis of DNA cloned in filamentous phage: influence of hemimethylated GATC sites on marker recovery from restriction fragments. *Nucl. Acids Res.*, **10**, 6475–500.
56. Radman, M., Wagner, R. E., Glickman, W. and Meselson, M. (1980) in *Developments in Toxicology and Environmental Sciences*, Vol. 7 (ed. M. Alacevic), Elsevier, Amsterdam, p. 130.
57. Winter, G., Fersht, A. R., Wilkinson, A. J., Zoller, M. and Smith, M. (1982) Redesigning enzyme structure by site directed mutagenesis: tyrosyl tRNA synthetase and ATP binding. *Nature*, **299**, 756–8. (A lucid account of the structural features of enzymes important for catalytic function is to be found in Fersht, A. R. (1977) *Enzymes Structure and Mechanism*, W. H. Freeman, San Francisco.)

4 Bacterial plasmid vectors

Plasmids are extrachromosomal replicons which fall into two major categories depending upon whether or not they can promote their own transfer from one bacterial strain to another through a conjugative process. The archetypal example of a conjugative plasmid is the F factor of *E. coli*. The transfer of DNA from a male strain of *E. coli* carrying the F genome to a female strain which lacks the F genome, requires the products of about 20 transfer (*tra*) genes which are plasmid encoded [1]. In this chapter we will see examples of both conjugative and non-conjugative plasmids that have been developed as cloning vectors. For the main part, however, the cloning vectors that have been developed are derived from non-conjugative plasmids. This is largely because such vectors were constructed at a time when there was considerable uncertainty about the hypothetical hazards of allowing DNA to cross species barriers as a result of laboratory experimentation. The fear was that a potentially dangerous gene could be accidentally transferred to a variety of bacteria if cloned upon a vector capable of conjugation. The conjugative plasmids are capable, not only of their own transmission from one cell to another, but they can also transfer chromosomal markers. This occurs at a high frequency if the conjugal plasmid becomes integrated into the bacterial chromosome. However, conjugal plasmids can also mobilize DNA to which they are not covalently joined, including coexisting non-conjugative plasmids. The non-conjugative plasmids which have been developed as vectors have been modified to give more effective cloning vehicles which can no longer be effectively mobilized during conjugation. A comprehensive background to the molecular biology of plasmids can be found elsewhere [2].

In addition a number of other properties are desirable in a plasmid cloning vector: it should have a replicon capable of generating a high copy number of recombinant plasmids per cell; it should have a marker function so that bacteria carrying the plasmid can be directly selected and it should have a variety of cleavage sites for restriction endonucleases commonly used for cloning. These sites should be present once in the genome in non-essential regions. The trend has been to minimize the size of the vectors in order to remove genes involved in transposition or conjugation and also to remove non-essential DNA containing duplicated restriction cleavage sites. This chapter will first trace the development of plasmid vectors for *E. coli*, that workhorse of the molecular biologist, and then show how the same principles are being applied to plasmid vector systems of other microorganisms.

4.1 Vectors for *E. coli*

4.1.1 pSC101 A low copy number plasmid

pSC101 was the first effective cloning vehicle to be used for cloning eukaryotic DNA [3]. In these experiments *Eco*RI fragments carrying the genes for ribosomal RNA from *Xenopus laevis* were introduced into the pSC101 replicon. At this time the ideas for cloning DNA in plasmids had been around for a couple of years. Schemes had been put forward for constructing recombinants using a plasmid, λdv*gal*, derived from a defective transducing phage [4]. This work had not gone ahead since it would have involved introducing a recombinant λdvgal/SV40 molecule into *E. coli* and it was uncertain at that time whether this had any hazardous implications. The practical application of λdv*gal* as a vector was, in any case, limited by the technical approaches in use at that time. Agarose gel electrophoresis was not then used for DNA fractionation and only a limited number of restriction endonucleases had been discovered. *Eco*RI was one of the few restriction enzymes available and this cleaves λdv*gal* within the *O*, *P* region thereby necessitating the use of a dimer molecule which would have one intact *O*, *P* region and therefore be capable of DNA replication. pSC101, on the other hand, not only has the advantage of a single *Eco*RI site at which DNA can be inserted, but also carries a strong selectable marker for tetracycline resistance. Neither of these functions are affected by the insertion of foreign DNA at the *Eco*RI site. The 9 kb plasmid has the disadvantage of being under stringent replicative control and is present in only 1–2 copies per cell. Consequently the yields of plasmid DNA from cells carrying pSC101 are low by comparison with the plasmid vectors that are in current use.

pSC101 is derived from the conjugative plasmid R6–5. This R factor shares about 50 kb homology with F in that region which includes the *tra* operon, those genes necessary for the conjugal transfer of DNA [5], and carries the genes for resistance to several antibiotics. pSC101 arose from an experiment in which R6–5 DNA was hydrodynamically sheared and then used to transform *E. coli* to tetracycline resistance [6]. It is possible that it arose from the cyclization of a segment of R6–5 DNA and the concomitant activation of a tetracycline resistance gene which is not active in the parental plasmid. Alternatively, pSC101 might have coexisted with the R factor as a small, sheer resistant, supercoiled DNA molecule.

4.1.2 ColEl

The problem of low plasmid yield does not occur with the ColEl plasmid, which is present in about 20 copies per cell. This can be increased to between 1000 and 3000 copies per cell by addition of chloramphenicol to a log-phase culture. Under these conditions the chromosomal DNA stops replicating, whereas the plasmid DNA continues to replicate and eventually constitutes about 50% of the cellular DNA [7]. ColEl is one of a series of plasmids which determine the production of the antibiotic proteins, the colicins. Each colicin has a characteristic mode of action: colicins E1 and K inhibit active transport [8, 9]; colicin E3 inhibits protein synthesis by cleaving 16S rRNA [10] and colicin E2 promotes the degradation of DNA [11]. In all cases the end result is cell death unless that cell carries the colicinogenic plasmid. Cells carrying a Col plasmid are immune to the effects of the colicin specified by that plasmid. This immunity forms the basis of a selection system for cells transformed by Col plasmids. The selection system has to be applied with care, however, since cells resistant to colicins arise spontaneously at quite a high frequency in a bacterial population. In spite of these difficulties the high copy number of ColEl plasmids is advantageous and ColEl has been used extensively as a cloning vehicle [12]. Foreign DNA may be inserted into the single *Eco*RI site. This renders the plasmid unable to produce colicin, but neither the replication properties nor the ability to confer colicin immunity are affected.

4.1.3 Derivatives of ColEl which contain drug resistance markers

In order to improve methods for selecting transformants, drug resistance genes have been introduced into ColEl or its derivatives. This has been achieved either by joining restriction fragments *in vitro* or by selecting plasmids which have acquired drug resistance markers following transposition *in vivo*. The development of the commonly used plasmid vectors has followed the tortuous pathway summarized in Fig. 4.1. One of the first derivatives of ColEl to carry a drug resistance marker, pML2, was an *in vitro* recombinant between

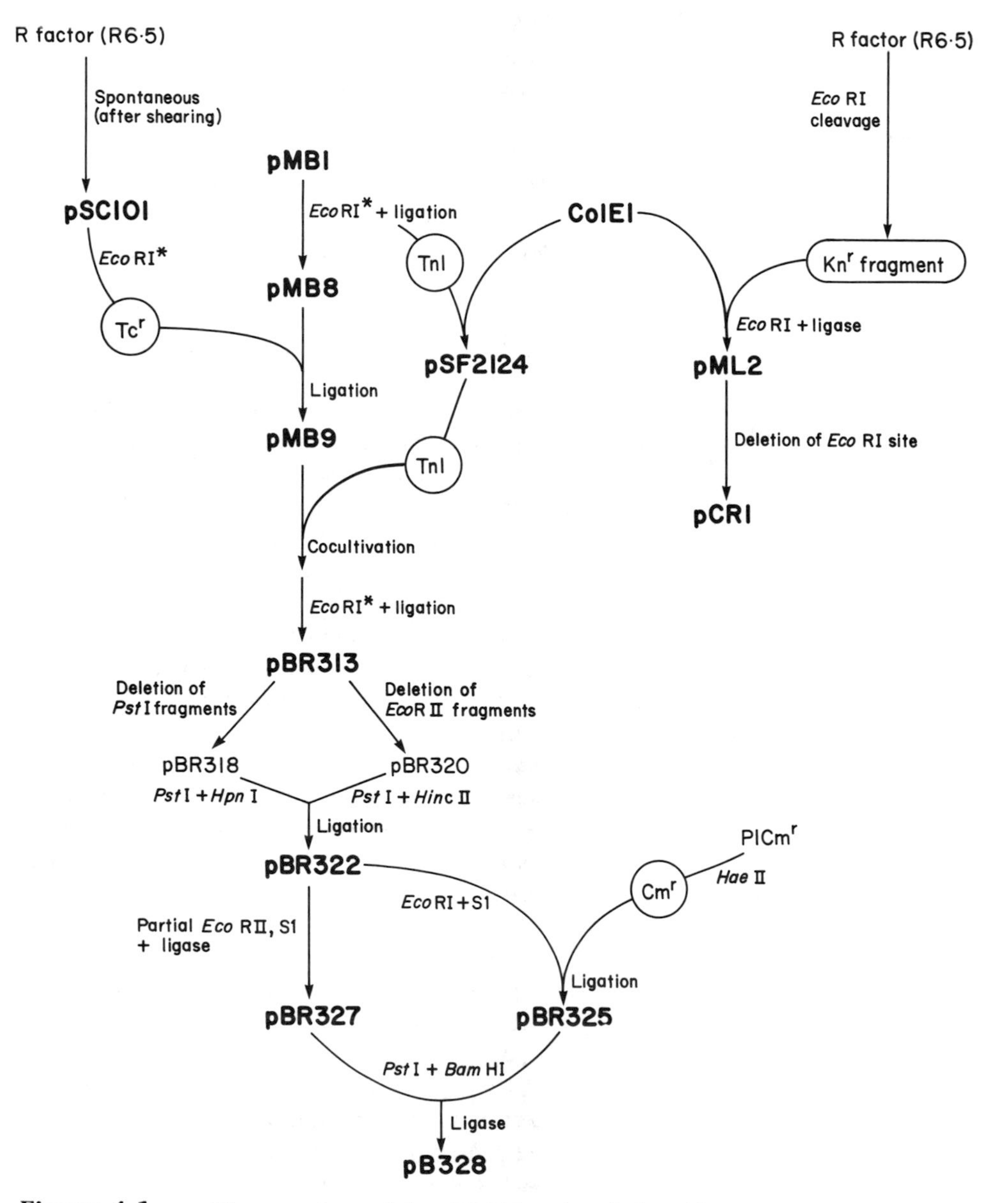

Figure 4.1 The geneology of the pBR312 series of plasmid vectors.

*Eco*RI-cleaved ColEl and an *Eco*RI fragment which carries the gene for kanamycin resistance [12]. pML2 therefore has two cleavage sites for *Eco*RI and so it is difficult to use as a vector for *Eco*RI fragments. This problem was overcome in pCR1 by deleting sequences around one of these *Eco*RI sites *in vitro* using one of the techniques described in Chapter 2 [13]. The gene for β-lactamase which determines resistance to the antibiotic ampicillin was introduced into ColEl by the translocation *in vivo* of the transposon Tn1 (formerly TnA). This was carried out by cocultivating an R plasmid (R1 drd 19) which carries Tn1 together with ColEl in the same bacterial cells [14]. One of the recombinants, pSF2124, that specifies both colicin biosynthesis and ampicillin resistance, has been used as a cloning vector in its own right but has also been incorporated into other vectors. pSF2124 has single sites for *Bam*HI and *Eco*RI into which foreign DNA may be inserted without affecting the ampicillin resistance marker.

In the subsequent development of plasmid vectors, the trend has been to reduce the size of the plasmid genome. This removes segments of the plasmid genome that are non-essential for its role as a vector and thereby eliminates unnecessary restriction endonuclease cleavage sites. The intention is furthermore to incapacitate any transposon within the plasmid. Transposons are capable of *rec*A independent translocation from one replicon to another. This process often generates deletions which extend from within the transposon into the externally flanking sequences. For any such events to occur within a cloning vector is clearly disadvantageous, since they could result in the loss of selectable marker and the potential loss or even rearrangement of the cloned DNA sequence. The transfer of DNA from one replicon to another is also undesirable since this would offer a route whereby potentially hazardous genes could 'escape' from the laboratory environment. Deletions within the repetitive elements which flank transposons eliminate the ability of the element to transpose. During the course of plasmid diminution during vector development, segments of these terminal repeated sequences have been deleted and the drug resistance markers are no longer capable of transposition.

The replicon present in most of the plasmid vectors that are now in common use does not originate from ColEl but from the extremely closely related plasmid pMB1. This plasmid carries genes that determine resistance to ampicillin, and also the genes for the *Eco*RI restriction–modification system. A set of experiments were performed in order to reduce this plasmid in size but to leave its replication origin and a selectable marker, immunity to colicin E1. This was carried out by digesting pMB1 with *Eco*RI under conditions in which the enzyme recognizes four of the nucleotides of the *Eco*RI recognition site (*Eco*RI* conditions), and so cleaves at additional sites within the genome. The resulting fragments with *Eco*RI AATT cohesive termini can be religated to give circular molecules. When these are introduced into *E. coli*, those which contain the plasmid replication origin will successfully transform the bacterial cells. It is therefore possible to select recombinants which have lost *Eco*RI* fragments carrying markers from the original plasmid. One such colicin E1 immune ampicillin-sensitive clone contained a 2.7 kb plasmid with a single *Eco*RI site which has been designated pMB8 [15].

DNA segments carrying drug resistance genes were subsequently introduced into the plasmid pMB8. The first such fragment was derived from pSC101 and contained the tetracycline resistance gene. This was obtained by digesting pSC101 with *Eco*RI under *Eco*RI* conditions. The *Eco*RI* fragments were ligated to pMB8 that had been cleaved with *Eco*RI. One of the recombinant plasmids arising from this experiment was the 5.3 kb pMB9 which combines the advantageous replicative properties of ColEl with the Tc^r marker. This plasmid has a single *Eco*RI site and has been extensively used as a vector. pMB9 also has single sites for *Bam*HI and *Sal*I, but insertion of foreign DNA into these sites inactivates the Tc^r marker. It also has a single site for *Hin*dIII which is in the promoter region of the tetracycline resistance gene. The plasmid can be protected from *Hin*dIII cleavage if RNA polymerase is bound to it. Insertion of DNA into this site inactivates the tetracycline resistance gene unless the foreign DNA provides a sequence which can restore promoter activity. Quite fortuitously, this appears to be the case for the *Hin*dIII fragments of the 5S ribosomal RNA genes of *Xenopus* which permit the expression of tetracycline resistance when cloned into this site.

In order to be able to use all the unique sites in the tetracycline resistance gene for cloning and yet still be able to apply strong selection, the gene for ampicillin resistance (Ap^r) has been introduced into pMB9. This was achieved by cultivating pMB9 with pSF2124 so that Tn1 could undergo transposition from one plasmid onto the other. A number of plasmids which determined resistance to both antibiotics were isolated from such experiments. The Tn1 transposon contains a site for *Bam*HI and so this set of plasmids no longer had a single site for this enzyme. In order to remove this additional *Bam*HI site, one of the plasmids was subjected to *Eco*RI* digestion and religation of the resulting fragments. Again the DNA was introduced into *E. coli*, and transformants selected on tetracycline and ampicillin. One of the transformants harboured the plasmid pBR 313, which has only the single *Bam*HI site of the Tc^r gene. Foreign DNA can be inserted into this site or the *Hin*dIII and *Sal*I sites, thereby inactivating the Tc^r gene but leaving the Ap^r gene functional (see Fig. 4.2). The deletion of sequences containing the *Bam*HI site of Tn1 also prevents translocation of the Ap^r onto other episomes [16].

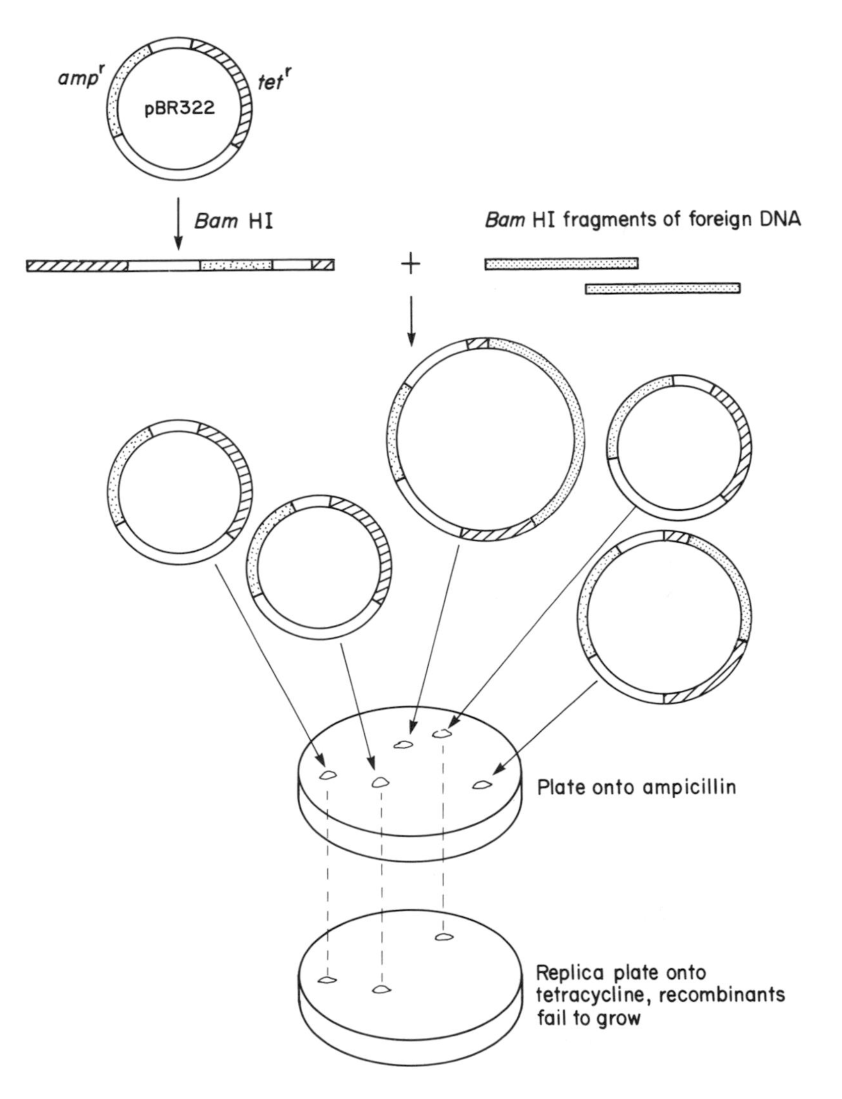

Figure 4.2 Screening for recombinants in plasmids carrying two drug resistance markers.

In a final set of experiments two *Pst*I sites were eliminated from pBR313. This leaves a plasmid with a single *Pst*I site in the β-lactamase gene. When DNA is cloned into this site, the β-lactamase gene is inactivated resulting in recombinant plasmids which have an Ap^sTc^r phenotype. This was achieved by recombining two restriction fragments from pBR313 *in vitro* [17]. The resulting plasmid was pBR322, one of the most extensively used cloning vectors.

The gene for chloramphenicol resistance (Cm^r) has also been introduced into pBR322. This gene is carried on a *Hae*II fragment from a P1-*Cm* transducing phage. It was ligated to *Eco*RI-generated linears of pBR322 which had been treated with S1 nuclease to remove the *Eco*RI cohesive ends. The ligation of these fragments which have fully base paired ends effectively destroys the *Eco*RI sites (see Chapter 2) and the only remaining *Eco*RI site is located in the Cm^r gene. Inactivation of the Cm^r gene of this plasmid, pBR325, can be used to identify *in vitro* recombinant plasmids in which the foreign DNA has been inserted into the *Eco*RI site [18].

4.1.4 Additional effects of genome diminution

The empirical approach to vector design involving a progressive reduction of genome size has had a number of desirable side effects. We have already seen examples of how transposons within vectors have been rendered incapable of translocation whilst leaving their drug resistance genes functional. It is also undesirable that a cloning vector can be conjugally transferred from one bacterium to another. This could lead to the spread of genes which specify hazardous products between bacterial species if the host were to escape from the laboratory environment. The parental plasmids from which the plasmid vectors have been derived are by themselves incapable of transfer, but they can be mobilized by transmissible plasmids. ColEl can, for example, be mobilized by F plasmids. ColEl contributes a mobility protein to this process, which is encoded by the gene *mob*. The *mob* gene has been lost during the creation of pBR322 which consequently cannot be mobilized by a transmissible plasmid [17]. pBR322 can, however, be mobilized if a third plasmid, for example ColK, is present [19]. It is proposed that in this case the ColK plasmid provides a 'mobility protein' which interacts in trans with a specific DNA sequence (the *nic* site) on pBR322. Although the gene for the 'mobility protein' has been deleted from pBR322, its site of action is still present. Several plasmids have been derived from pBR322 in which restriction fragments containing the *nic* site have been deleted. pBR327 and pAT153 are examples of such plasmids which still retain genes specifying resistance to tetracycline and ampicillin [20]. The *Pst*I/*Bam*HI fragment from pBR327 which contains this deletion has been ligated to the *Pst*I/*Bam*HI fragment of pBR325 which carries the Cm^r gene to produce a recombinant pBR328. The plasmid pBR328, like pBR325, carries $Cm^rAP^rTc^r$ genes but now has additional single sites for *Pvu*II and *Bal*I.

Many of the essential functions of the commonly used plasmid vectors have now been tightly mapped on their genomes. A 580 base pair fragment of pMB1 has been identified which contains all of the information necessary for plasmid replication [21]. Additional sequences are, however, required to ensure stable inheritance of a plasmid, and its segregation into the daughter cells upon cell division. Meacock and Cohen [22] have identified a locus of pSC101 responsible for partition of plasmids between daughter cells which they designate *par*. The *par* locus lies within a 270 base pair fragment adjacent to the replication origin. Similar sequences are present in ColEl but are not yet as well defined. It is probable that many cloning vectors are *par*$^-$ as a consequence of the steps taken to reduce their size. This could create a problem with industrial scale cultures. When cells containing pBR322 or PMB9 are grown in a chemostat with a limiting supply of nutrient, then plasmid free segregants are found to arise [23]. The solution would be either to grow the culture under continuous selection or to incorporate the *par* locus into the vector.

4.1.5 High copy number plasmids that permit histochemical screening for recombinants

The principle of insertional mutagenesis as a means of identifying plasmid recombinants was established with plasmids such as pBR322 (see Fig. 4.3). The identification of recombinants does, however, necessitate replica plating in order to recognize the loss of phenotype. Vectors have therefore been developed in a number of laboratories to exploit the highly sensitive screen for functional β-galactosidase using media containing its chromogenic substrate X-gal. The insertion of foreign DNA inactivating the *lacZ* gene in these vectors can be recognized directly upon transfection of recombinant DNA molecules into *E. coli*. Rather than detail all the vectors that have been constructed which use this principle, this discussion will concentrate upon only one such set of vectors. These were developed by Vieira and Messing and use the same segment of the *lac* operon as that in the Ml3 cloning vectors which have already been discussed in Chapter 3 [24]. It encodes the *N*-terminal α-peptide of β-galactosidase which is capable of the intra-allelic complementation of *N*-terminal defective proteins encoded by the host. This complementation results in functional β-galactosidase activity recognized by the production of blue bacterial colonies when cells are plated on media containing X-gal and the *lac* inducer, IPTG. The segment of the *lac* operon was transferred from the Ml3 cloning vector, Ml3mp7, onto a segment of pBR322 lacking the gene for tetracycline resistance. In order to make full use of the cloning sites in the *lacZ* gene, the segment of pBR322 DNA had first to be modified to remove cleavage sites. This involved mutagenesis *in vivo* with ethylmethane-sulphanate and hydroxylamine to remove *Pst*I and *Hinc*II sites in a manner analogous to that which had been applied in the construction of the Ml3 vectors. Furthermore an *Acc*I site was removed by generating deletion mutants using *Bal*31 *in vitro*. The resulting

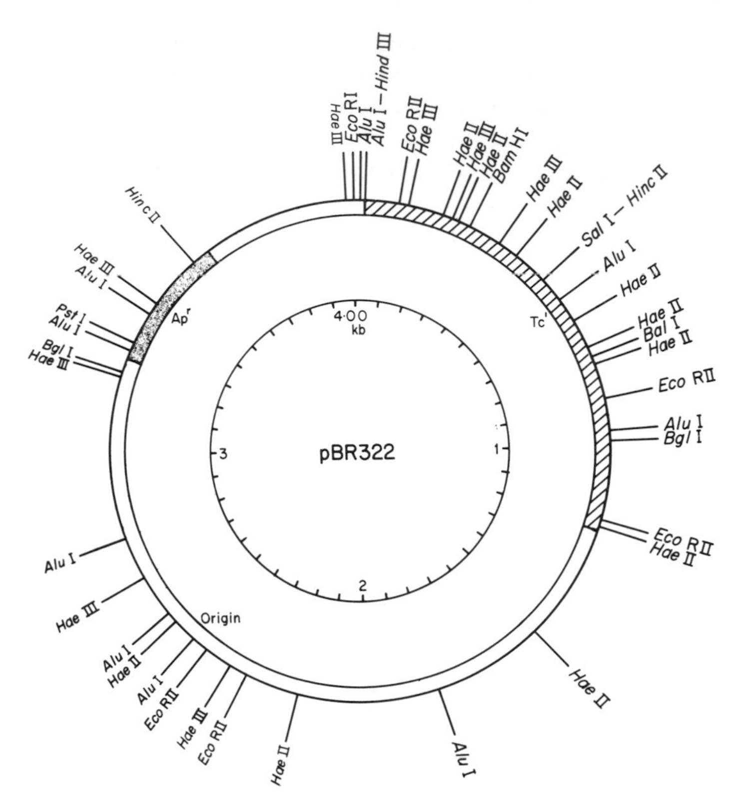

Figure 4.3 The physical map of pBR322.

vector, pUC7, has unique sites for these enzymes in the 5′ end of the *lac*Z gene (see Fig. 4.4). A further two vectors, pUC8 and pUC9, have been constructed which have a different multiple restriction enzyme cloning site in opposite orientations analogous to the Ml3 vectors, Ml3mp8 and Ml3mp9. The multiple restriction enzyme cloning site

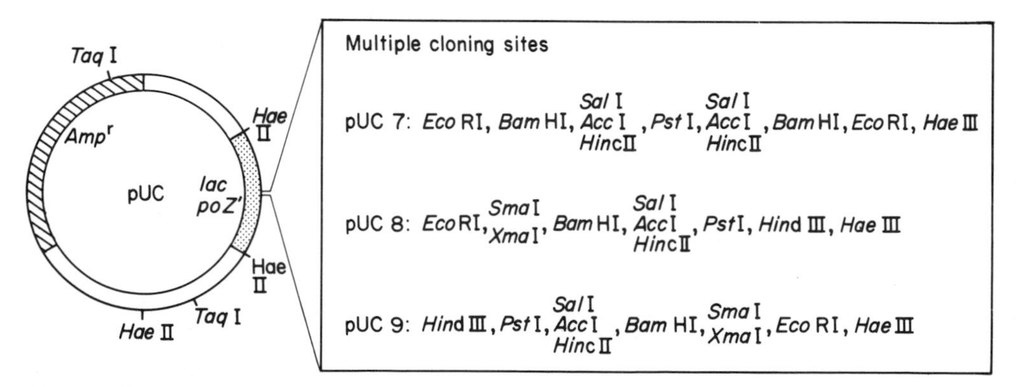

Figure 4.4 The multiple cloning sites of the pUC plasmids.

maintains the reading frame of β-galactosidase in each of these vectors. Insertions of DNA generally disrupt the reading frame resulting in the loss of activity of the α-peptide and the production of white colonies. The exceptions are insertions which maintain an open reading frame in the recombinant DNA molecule and which are capable of producing functional α-peptide. The detection system has now been turned around in order to develop vectors in which one can screen specifically for DNA segments with open reading frames.

4.1.6 Open reading frame vectors

We have seen previously how gene inactivation by the insertion of a foreign piece of DNA has been used as a means of identifying recombinants. A gene may not necessarily be inactivated if the insertion is in a position such that the function of a hybrid protein would not be affected if the translational reading frame is maintained. A number of laboratories have directed their attention towards the construction of vectors that can direct the synthesis of such hybrid proteins. These can be used to screen for cloned DNA segments containing open reading frames. This has been achieved by modifying the cloning vector so that it contains an inactive marker which is out of translational phase. The activity of the marker gene is restored if a segment of DNA containing an open reading frame is inserted into the plasmid in correct translational phase (Fig. 4.5).

In three such vectors the body and 3′ end of the *lacZ* gene has been fused to the promoter and 5′ end of some other gene such that it is out of phase and cannot produce functional β-galactosidase. Cloning

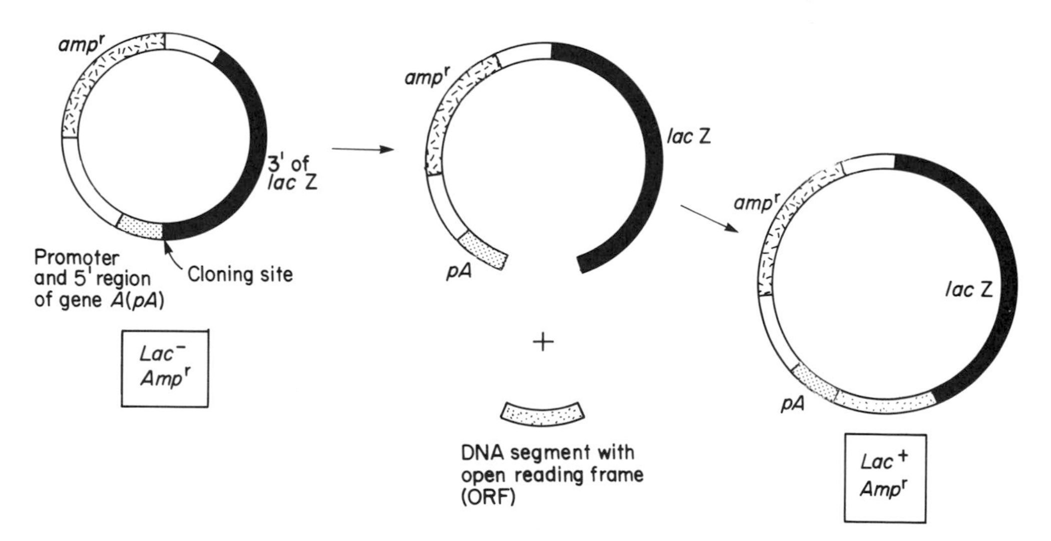

Figure 4.5 Cloning DNA in open reading frame vectors.

sites have been placed at the point of this gene fusion which permit the insertion of foreign DNA. If the inserted DNA restores the translational reading frame then a tripartite protein can be produced with functional β-galactosidase activity. These vectors are shown in Fig. 4.6. The vector constructed by Gray *et al.* [25] (see Fig. 4.6(a) . . . pMR100) has a fusion of (a) the *cI* gene under the control of the *lac* promoter with (b) a fused *lac*I–Z gene. These two segments were originally joined at a *Hin*dIII site, but a *Bam*/*Sma*I

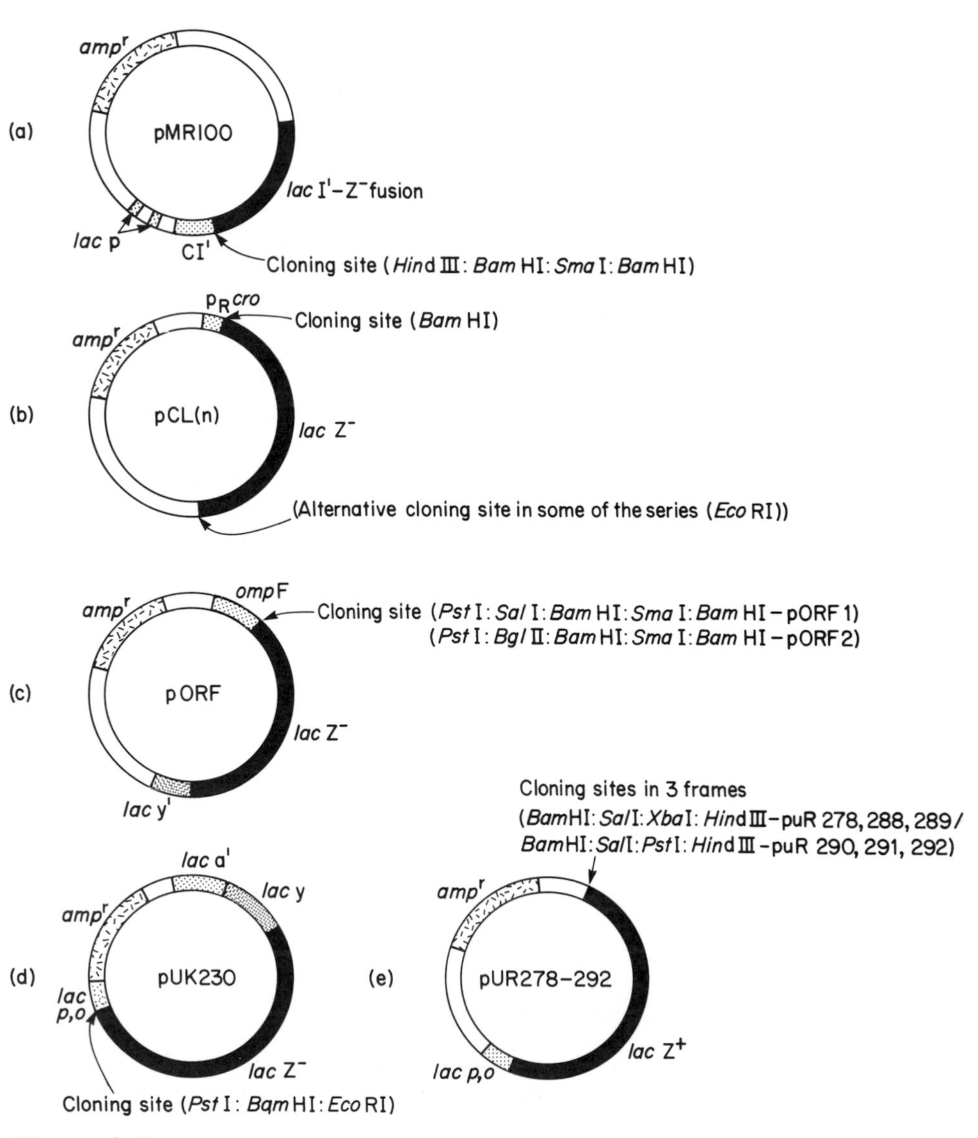

Figure 4.6 A selection of open reading frame vectors.

linker was subsequently inserted in order to destroy the reading frame. Zabeau and Stanley [26] have constructed a series of vectors in which the beginning of the *cro* gene (under the transcriptional control of p^R) is fused to the *lacZ* gene (Fig. 4.6(b) . . . pCLn). DNA can be inserted at the *Bam* site between the *cro and lac* genes and can potentially restore an open reading frame. Weinstock *et al.* [27] have utilized a segment of the *ompF* gene of *E. coli*, which encodes an outer membrane protein (Fig. 4.6(c) . . . pORF). The segment of the *ompF* gene provides the signal sequences at the *N*-terminus of the membrane protein. Any hybrid peptides produced by this vector are therefore transported from the cytoplasm to sites at which they can readily be detected by the immunological screening techniques. This element is fused with the *lacZ* gene and the 5′ end of *lacY* with a polylinker containing sites for *PstI*, *BglII*, *BamHI*, and *SmaI* in such a way as to destroy the reading frame with respect to β-galactosidase. The overproduction of a hybrid protein with β-galactosidase activity in this case results in a lethal phenotype. This has also been observed by Zabeau and Stanley [26] and is possibly due to a sequence downstream from the *lacZ* gene. The lethality can be prevented by mutations in *ompR*, a positive regulator of *ompF*.

Koenen *et al.* [28] have designed a vector to monitor the insertion of open reading frames in fusion proteins solely with β-galactosidase. Their vector is derived from pUK217 [29], a vector similar to the pUL series in that it contains a polylinker sequence at the same site at the 5′ end of the *lacZ* gene. Unlike the pUC plasmids, however, it contains the entire *lacZ* gene together with the *lacY* and part of the *lacA* gene. The plasmid pUK230 (Fig. 4.6(d)) has a 7 base pair deletion in the polylinker region that generates a frame shift. Insertion of foreign DNA has the potential of restoring β-galactosidase activity. In a second series of similar vectors, multiple cloning sites have been created at the 3′ end of the *lacZ* gene [30]. The series provides these cloning sites in all three translational frames (Fig. 4.6(e)). These plasmids are capable of directing the synthesis of β-galactosidase fused in its *C*-terminal region to a foreign peptide. This may have considerable advantages since Stanley [31] has reported that similar constructs produce fused proteins which have increased stability.

4.2 Vectors for other Gram negative organisms

4.2.1 Broad host range vectors

The plasmid vectors which have been described in the preceding part of this chapter can only be propagated in *E. coli* and closely related enteric bacteria. There are many non-enteric Gram negative bacteria of considerable commercial interest which will not serve as hosts for these vectors. These include strains of *Pseudomonas* which can degrade a variety of organic chemicals and the nitrogen fixing strains of *Rhizobium* and *Azotobacter*. Fortunately, there are a number of naturally occurring plasmids which have a broad host range and are capable of propagation in all of these species. Steps have been taken to develop cloning vectors from some of these plasmids. Most of this work has been carried out on RSF1010 or on two related plasmids RP4 and RK2. These are members of different incompatibility groups: RSF1010 is from the *Q* group; RP4 and RK2 are from the *P*-group. The two groups cannot coexist in the same cell. The broad host range plasmids have very few cleavage sites for restriction enzymes with hexanucleotide recognition sequences. Presumably this is a trait that has been selected during their evolution in a variety of bacterial host species carrying different restriction systems. Most of the restriction sites which are useful for cloning are carried in their drug resistance markers, which have probably been acquired more recently in the evolution of the plasmids, by transposition. A second difficulty in their development as vectors is that genes for replication and conjugal transfer are dispersed over their genomes.

RP4 is a 56 kb plasmid which encodes resistance to ampicillin, kanamycin and tetracycline. It has been used directly as a cloning vehicle in cloning DNA from *Rhizobium* and *Proteus*, for example [32]. Smaller derivatives have, however, been constructed by partial *Hae*II digestion of the related plasmid RK2. One of these, pRK248, is 9 kb in size and specifies tetracycline resistance. Another, pRK2501, has an additional 2 kb *Hae*II fragment giving kanamycin resistance [33]. The kanamycin resistance gene has sites for *Hin*dIII and *Xho*I and the tetracycline resistance gene has a *Sal*I site. In addition pRK2501 has sites for *Eco*RI and *Bgl*II into which DNA can be cloned without any phenotypic consequence. The main property required of a vector derived from a broad host range plasmid is that it should allow the introduction of a given gene into a variety of species. This criterion is quite the opposite of those which operated in the development of *E. coli* vectors, where because of worries about the potential hazards of unknown genes, the aim was to restrict the recombinant molecule to the *E. coli* cell. In this light some broad host range vectors have the problem of being non-mobilizable and so they are effectively restricted to use with strains that can be readily transformed with DNA. This problem has been neatly circumvented by Ditta *et al.* [34] who have devised a tripartite cloning system from the RK2 plasmid. This consists of two plasmids: the first is a non-mobilizable low molecular weight derivative of RK2, pRK290, which carries the tetracycline resistance marker. This can be propagated in *E. coli* and is used as the cloning vector *per se*. The second consists of a ColEl replicon carrying the RK2 transfer genes, which because of the narrow host range of ColEl is confined to *E. coli* strains. DNA cloned in pRK290 can be transferred to another recipient Gram negative bacterium in a triparental mating. The ColEl type plasmid carrying the transfer genes is transferred into the *E. coli* strain carrying the pRK290 recombinant and thereby causes the transfer of this plasmid into the third bacterium of another species.

RSF1010 is an 8.9 kb plasmid which carries determinants for sulphonamide resistance and streptomycin resistance. It has unique sites for *Eco*RI, *Hpa*I, *Pvu*II, *Sst*I and *Bst*II and will transform strains of *Pseudomonas* with a moderately high efficiency. An early problem was encountered in the restriction system of *Pseudomonas* species which forms an effective barrier to the introduction of foreign DNA into cells by transformation. Strains of *P. aeruginosa* and *P. putida* which are defective in their restriction system have now been described [35]. These should be highly suitable hosts for recombinant DNA molecules. RSF1010 has also been modified by the introduction of alternative drug resistance markers and restriction sites suitable for cloning. RSF1010 has two *Pst*I sites which flank the sulphonamide resistance gene. This restriction fragment has been replaced with *Pst*I fragment carrying chloramphenicol resistance determinants from different sources or with *Pst* fragments endowing both kanamycin and sulphonamide resistance. Some of these plasmids had their mobilization genes deleted using *Bal*31 nuclease (see Chapter 2). The development of RSF1010 as a cloning vector has therefore followed a set of steps which exactly parallel those taken in the development of plasmid vectors for *E. coli*.

4.2.2 Plasmids of *Pseudomonas*

Pseudomonas species contain a variety of plasmids which encode catabolic enzymes capable of degrading an enormous range of natural and synthetic organic compounds. Of these, perhaps the most intensively studied is a plasmid pWWO from *P. putida*. This is one member of a set of plasmids which have required the generic name TOL plasmids. It is 117 kb in length and carries genes which permit the utilization of toluene, *m*- and *p*-xylene, 3-ethyltoluene and 1,2,4-trimethylbenzene as well as their alcohol, aldehyde and carboxylic acid derivatives (Fig. 4.7). The plasmid can be conjugally transferred into *E. coli*. This has facilitated the mapping of the genes of the degradative pathway, largely by transposon mutagenesis [36]. The pWWO plasmid is propagated in a strain of *E. coli* carrying the transposon Tn5 on its chromosome. After 70 or so generations it is transferred back into *P. putida* by conjugation. Trans-conjugants of *P. putida* are selected which have acquired pWWO carrying Tn5 which confers resistance to kanamycin. The Tn5 transposon inactivates

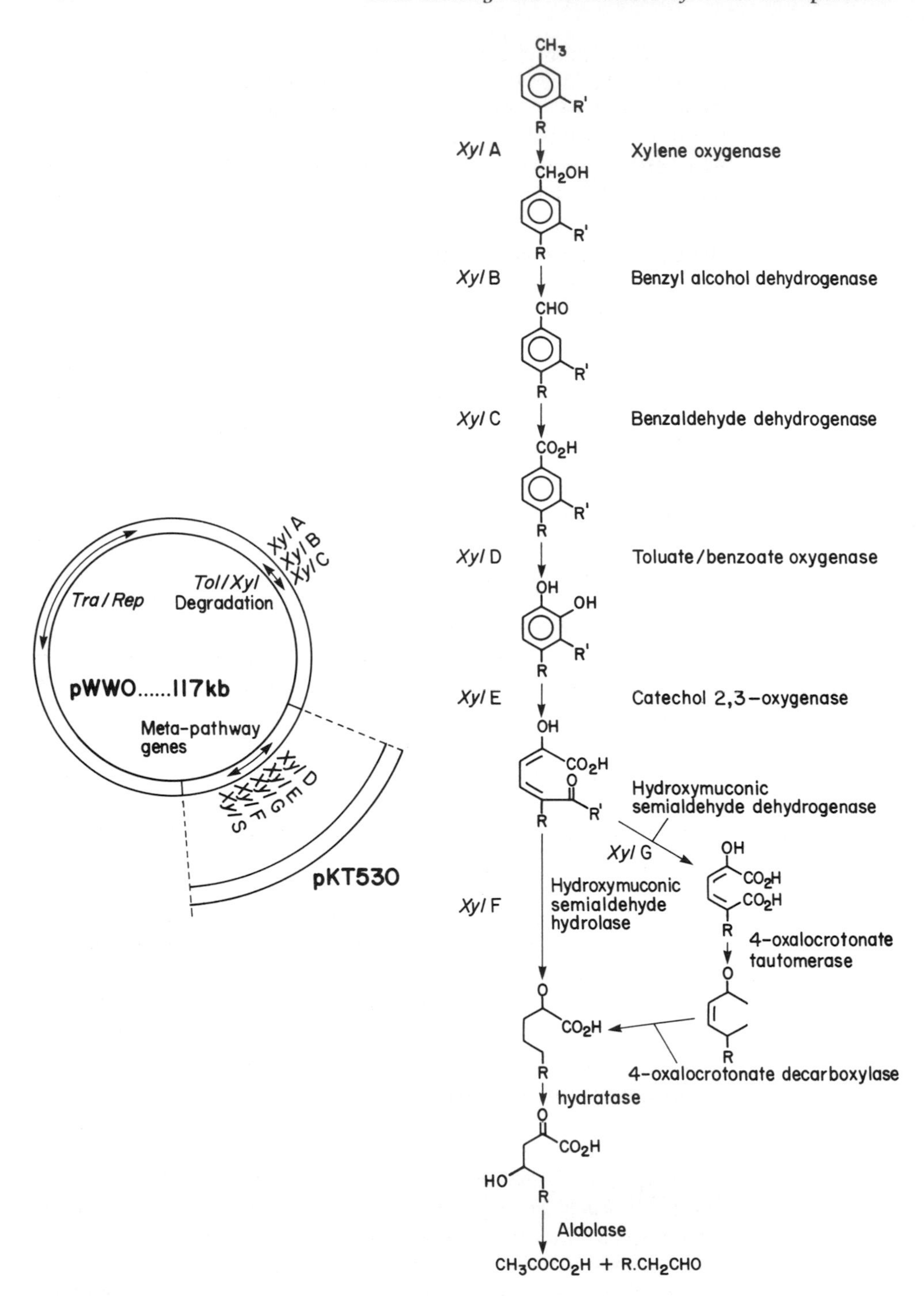

Figure 4.7 The physical map of pWWO and the pathway for toluene degradation.

those genes into which it is inserted. A gene tagged in this way can be mapped using one of the physical mapping techniques described in Chapter 6. There are two clusters of genes for the degradative pathways; one cluster encodes the enzymes which convert the hydrocarbon to the carboxylic acid (*xyl*A, *xyl*B, *xyl*C – the upper pathway) and the other encodes toluate oxidase and the so-called 'meta-cleavage' pathway enzymes (*xyl* D–F). These two segments of the catabolic pathway are thought to be controlled by at least two regulatory genes; *xyl*R which stimulates the expression of the genes of both pathways and *xyl*S which together with carboxylic acid substrates stimulates the expression of the meta-cleavage pathway [34]. The genes for the meta-cleavage pathway have been cloned on one of the broad host range vectors for Gram negative organisms derived from RSF1010 which were described above [34]. This recombinant plasmid, pKT530, also carries the *xyl*S gene in addition to the gene for the catabolic enzymes (Fig. 4.7). The *xyl*S gene encodes a transacting regulatory element, which must also interact with the product of the *xyl*R gene, since when *xyl*S mutants are grown in the presence of a hydrocarbon inducer, the upper pathway is induced but not the meta-cleavage pathway. The genes encoding the degradative enzymes carried on pKT530 are expressed in *E. coli* but at 1–5% of the level found in *P. putida*. High levels of expression can be achieved when the genes are placed under the control of an *E. coli* promoter [38].

A large number of TOL plasmids have been isolated, all of which appear to have a similar pattern of regulation. The meta-cleavage part of the pathway is common to them all. In some strains of *Pseudomonas* there is an alternative pathway for benzoate degradation which is chromosomally encoded. The meta-cleavage pathway responsible for phenol and cresol degradation in strains of *P. putida* and *P. aeruginosa*, for example, has been shown to be chromosomally encoded. The chromosomal pathway can be more efficient than that carried on the plasmid and in these cases the entire plasmid or segments carrying the catabolic genes can be lost [39, 40]. It is possible that the genes for the entire catabolic pathway are carried on a large transposable element. This would help explain the widespread occurrence of metabolic pathways in a variety of genera.

The organisms carrying these plasmids are of considerable importance for the degradation of toxic organic wastes within our environment. There is, however, a considerable problem regarding the breakdown of various halogenated compounds used in insecticides and herbicides. This is because there are only a limited number of naturally occurring compounds with halogen substitutions and correspondingly few enzymes capable of degrading them. Many simple chlorinated compounds are amenable to microbial degradation, by enzymes which in many cases are plasmid encoded [41, 42, 43]. Several groups have attempted to extend the substrate requirements of these bacteria by introducing additional plasmids into them. Chakrabarty's laboratory has, for example, described a plasmid pAC25 which determines a degradative pathway for 3-chlorobenzoic acid [41]. Introduction of pWWO into this strain of *P. cepacia* led to the isolation by recombination *in vivo* of variants which could also utilize 4-chlorobenzoic acid and 3,5-dichlorobenzoic acid. In such a case, the critical factor is probably the wide substrate-specificity of the toluate oxidase encoded by the TOL plasmid. The plasmids undergo considerable rearrangements during the selection of the *P. cepacia* strains on the new substrates. In one case a novel plasmid was formed which comprised an 8.5 kb fragment of the TOL plasmid containing the replication, incompatibility and copy number genes and a duplicated segment of pAC25 needed for chlorobenzoic acid degradation [44]. The degradative genes carried on a 45 kb segment of the TOL plasmid undergo chromosomal integration. Chakrabarty's group has used this general approach to develop, from a mixed culture of *P. ceparia*, a pure culture that can utilize 2,4,5–trichlorophenoxyacetic acid (2,4,5-T) as its sole source of carbon [45]. 2,4,5-T is highly toxic and was a major component of 'agent orange' used in the Vietnam war.

The strain developed by Chakrabarty's laboratory is effective in degrading 2,4,5-T from soil samples, and is a triumph for this approach of laboratory 'breeding' with naturally occurring plasmids. At the same time it is important that the molecular cloning of genes from these plasmids should continue using vectors such as those derived from RSF1010. This will lead to a better molecular understanding of the control of these catabolic pathways and thereby permit the *design* of novel plasmids capable of the efficient degradation of other toxic wastes.

4.3 Vectors for Gram positive organisms

4.3.1 Plasmid vectors for *Bacillus subtilis*

The *Bacillus* species have attracted interest as cloning vehicles since they are Gram positive organisms which do not cause disease in humans. Furthermore their ability to sporulate has subjected them to scrutiny by molecular biologists. A number of methods have been developed for the transformation of *B. subtilis* following the demonstration that this species could be inefficiently transformed by supercoiled plasmid DNA [46]. The efficiency of transformation can be increased by using multimeric forms of the plasmid [47]. An alternative method of efficient transformation has been achieved using a linear recombinant DNA molecule transformed into a host cell already carrying a plasmid homologous to the vector [48]. This system requires recombination to occur *in vivo* between the in-going linear molecule and the resident plasmid, and is dependent upon the *B. subtilis recE* gene. A more practical system for efficient transformation by plasmids involves the uptake of DNA into protoplasts in polyethylene glycol [49].

Very few *Bacillus* plasmids have been discovered which are ideal for use as cloning vectors. Indeed most of the development of '*Bacillus*' cloning vectors has been carried out on plasmids isolated from *Staphylococcus aureus* which encode resistance to tetracycline (pT127) or chloramphenicol (pC194) [50]. A number of recombinant *Staphylococcus* plasmids have been constructed using approaches analogous to those described earlier in this chapter. These plasmids have two or more drug resistance markers and in some cases these are unique restriction sites within the genes which allow *in vitro* recombinants to be recognized by insertional marker inactivation. Undoubtedly useful *Bacillus* plasmids will be discovered. One such potentially useful plasmid, pAB124 encoding tetracyline resistance, has been isolated from the thermophile *B. stearothermophilus* [51]. This plasmid may be transformed into *B. subtilis*. Considerable interest is focussed upon thermophilic strains of bacteria as a means of producing enzymes suitable as catalysts for industrial processes operating at elevated temperatures. Enzymes from thermophiles have considerably greater resistance to thermal and chemical inactivation when compared to their mesophilic counterparts. We will undoubtedly see the development of thermophilic cloning systems for biotechnological applications.

The initial difficulties in establishing an efficient transformation system for *Bacillus* have led to the development of 'shuttle' vectors which have the capability of replicating both in *E. coli* and *B. subtilis*. Thus the primary cloning step may be carried out in *E. coli* and then once the recombinant has been isolated and characterized it may be transferred to *B. subtilis*. In its simplest form, this has been achieved by ligating the *S. aureus* plasmid pC194 to pBR322 to create a recombinant which in *E. coli* expresses chloramphenicol resistance and ampicillin resistance (the tetracycline resistance gene of pBR322 having been inactivated in joining the two plasmids together) [52]. The efficiency of the expression of bacterial genes in heterologous species varies considerably. The β-lactamase gene of the above recombinant is not expressed in *B. subtilis*. Studies using Northern blots (see Chapter 6) have failed to detect the β-lactamase mRNA. Chang and Cohen [53] found that the β-lactamase gene from *S. aureus* was expressed in *E. coli*, but on the other hand Brammar *et al.* [54] showed that the level of β-lactamase from *B. licheniformis* was about 600-fold lower when expressed in *E. coli* rather than in its normal host. This reflects dramatic differences in the control of both transcription and translation in different bacterial species. It is likely to be just as difficult to attain efficient expression of foreign prokaryotic genes in *E. coli* as foreign eukaryotic genes.

4.3.2 Plasmid vectors for *Streptomyces*

Streptomyces are Gram positive bacteria and so, like the *Bacilli*, they lack an outer membrane and produce numerous extracellular products, of which a number of antibiotics are of commercial interest. There is therefore considerable interest in the application of genetic engineering techniques to these species as a means of increasing the yields of the antibiotics. The life cycle of *Streptomyces* is shown in Fig. 4.8. The species are capable of sporulation and the

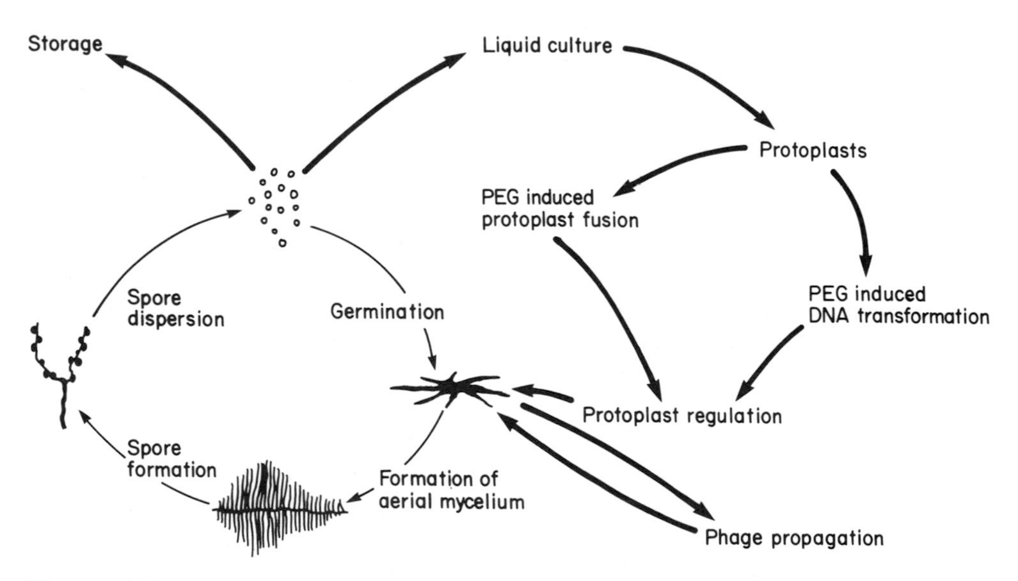

Figure 4.8 Introduction of DNA into *Streptomyces* at various stages in its life cycle.

spores germinate to grow as branching mycelia. *S. coelicolor* has an efficient sex plasmid, pIJ101 which has facilitated the genetic mapping of chromosomal markers. DNA can be introduced into protoplasts by a procedure analogous to that described for the protoplasts of *Bacilli*. This, and all other methods for the introduction of DNA into *Streptomyces* species, is dependent upon the effects of polyethylene glycol on the surface of the protoplasts. An alternative involves the entrapment of DNA in lipid vesicles. The conditions are similar to those used for the protoplast fusion of different *Streptomyces* species and both of these approaches are having considerable impact upon genetic studies of the species. DNA free liposomes can by themselves give a 100 fold stimulation of the uptake of DNA by protoplasts [55]. Most of the genetic studies have been carried out on *S. coelicolor*. It is *S. lividans* that has undergone the primary development as a cloning vehicle, however, since it can host a wide range of vectors.

Three plasmid cloning systems have been developed. Each plasmid type can undergo conjugal transfer, but cells which acquire the plasmids grow less well than when they lacked the plasmid. This phenotype is known as 'lethal zygosis' and is evident as a zone or 'pock' of cells in which growth is inhibited. These pocks are used as a means of assaying for cells bearing plasmids. Two of the 'parental' plasmids are low copy number plasmids with a narrow host range restricting their use to *S. lividans* and related species. One of the low copy number plasmids, SLP1.2, is one member of a set recovered in *S. lividans* following interspecific matings with *S. coelicolor*. The SPL1 replicon is integrated into the *S. coelicolor* chromosome and presumably prevents the extrachromosomal replication of homologous plasmids. SLP1.2 has a single *Bam*HI site and three *Pst*I sites in non-essential regions. These sites have been used for cloning antibiotic resistance determinants from other *Streptomyces* species: neomycin resistance from *S. fradiae*; thiostrepton resistance from *S. azureus*; and viomycin resistance from *S. vinaceus*. Several of these recombinants provide vectors with good selective markers; plJ61, for example, contains the thiostrepton resistance gene which can be used for selection and the neomycin resistance gene which provides a unique *Bam*HI site for insertional inactivation of the gene following the cloning of *Bam*HI fragments. The second low copy number plasmid is the 30 kb SCP2. Its use as a cloning vehicle has not been extensively developed. Its largest *Pst* I and *Bam*HI fragments have been used as vectors for shotgun cloning [56], but the lack of a plasmid stability function has inhibited their use. The third plasmid system is based upon the 8.9 kb plasmid pIJ101. It is usually present at about 100 copies per cell and has a broad host range within *Streptomyces*. Several derivatives have been made in which various plasmid functions are impaired, for example loss of the ability for independent conjugal transmission or modification of pock size. A variety of potential vectors have been constructed using restriction fragments carrying the drug resistance determinants described above. One vector uses a *Streptomyces* tyrosinase gene for clone recognition. This gene normally causes the production of brown pigment, whereas recombinants in which the gene has been inactivated by insertion of a restriction fragment specify white clones.

In passing it should be mentioned that a phage vector is also available for cloning in *Streptomyces* [57]. Phage vectors have been discussed in detail in Chapter 3. Suffice it to say here that the *Streptomyces* phage vectors based on phage ϕC31 are analogous to the bacteriophage λ vectors of *E. coli*. ϕC31 has a 42 kb linear DNA genome with cohesive ends. It lysogenizes many strains of *Streptomyces*. Derivatives of the phage having deletions have been isolated and recombinants have been constructed between such phage and pBR322. These 'shuttle' vectors have the useful property of being able to replicate as high copy number plasmids in *E. coli* or as phage in *Streptomyces*. The unique *Bam*HI and *Pst*I sites in the pBR322 moiety are attractive cloning sites.

The development of cloning systems in *Streptomyces* is clearly progressing rapidly. Expression systems have not yet been introduced, but this will no doubt occur once more knowledge is accrued of species specific promoters and ribosome binding sites. The pathway of vector development will probably parallel that taken in *E. coli*. The neomycin resistance gene of *S. fradiae* may become as attractive as the β-lactamase gene of pBR322, since it encodes the major soluble protein in cells carrying pIJ61.

References

1. Willetts, N. and Skurray, R. (1980) The conjugation system of F like plasmids. *Ann. Rev. Genet.*, **14**, 41–76.
2. Broda, P. (1979) *Plasmids*, W. H. Freeman, San Francisco.
3. Morrow, J. F., Cohen, S. N., Chang, A. C. Y. *et al.* Replication and transcription of eukaryotic DNA in *Escherichia coli*. *Proc. Natn Acad. Sci. USA*, **71**, 1743–7.
4. Jackson, D. A., Symons, R. M. and Berg, P. (1972) Biochemical method for inserting new genetic information into DNA of Simian virus 40 circular DNA molecules containing lambda phage genes and the galactose operon of *Escherichia coli*. *Proc. Natn Acad. Sci. USA*, **69**, 2904–9.
5. Sharp, P. A., Sohen, C. N. and Davidson, N. (1973) Electron microscope heteroduplex studies of sequence relations among plasmids of *Escherichia coli*. Structure of drug resistance (R) vectors and (F) vectors. *J. Mol. Biol.*, **75**, 235–55.
6. Cohen, S. N. and Chang, A. C. Y. (1973) Recircularisation and autonomous replication of a sheared R-vector DNA segment in *Escherichia coli* transformants. *Proc. Natn Acad. Sci. USA*, **70**, 1293–7.
7. Clewell, D. B. and Helsinki, D. R. (1972) Effect of growth conditions on the formation of the relaxation complex of supercoiled ColEl deoxyribonucleic acid and protein in *Escherichia coli*. *J. Bacteriol.*, **110**, 1135–46.
8. Fields, K. L. and Luria, S. E. (1969) Effects of colicins El and K on transport systems. *J. Bact.*, **97**, 57–63.
9. Fields, K. L. and Luria, S. E. (1969) Effects of colicins El and K on cellular metabolism. *J. Bact.*, **97**, 64–77.

10. Boon, T. (1972) Inactivation of ribosomes *in vitro* by colicin E3 and its mechanism of action. *Proc. Natn Acad. Sci USA*, **69**, 549–52.
11. Ringrose, P. (1970) Sedimentation analysis of DNA degradation products resulting from the action of colicin E2 on *Escherichia coli*. *Biochim. Biophys. Acta*, **213**, 320–34.
12. Hershfield, V., Boyer, H. W., Yanofsky, C., Lovett, M. A. and Helinski, D. R. (1974) Plasmid ColEl as a molecular vehicle for cloning and amplification of DNA. *Proc. Natn Acad. Sci. USA*, **71**, 3455–9.
13. Covey, E., Richardson, D. and Carbon, J. (1976) A method for the deletion of restriction sites in bacterial plasmid deoxyribonucleic acid. *Mol. Gen. Genet.*, **145**, 155–8.
14. So, M., Gill, R. and Falkow, S. (1975) The generation of a ColEl-Ap^r cloning vehicle which allows detection of inserted DNA. *Mol. Gen. Genet.*, **142**, 239–49.
15. Rodriguez, R. L., Bolivar, F., Goodman, H. M., Boyer, H. W. and Betlach, M. (1976) in *ICN–UCLA Symposium* 5 (eds. A. Danierlich *et al.*), Academic Press, New York, pp. 471–7.
16. Bolivar, F., Rodriguez, R. L., Betlach, M. C. and Boyer, H. W. (1977) Construction and characterisation of new cloning vehicles. Ampicillin resistant derivatives of the plasmid pMB9. *Gene*, **2**, 75–93.
17. Bolivar, F., Rodriguez, R. L., Greene, P. J. *et al.* (1977) Construction and characterisation of new cloning vehicles. A multi purpose cloning system. *Gene*, **2**, 95–113.
18. Bolivar, F. (1978) Construction and characterisation of new cloning vehicles. Derivatives of plasmid pBR322 carrying unique *Eco*RI sites for selection of *Eco*RI generated recombinant DNA molecules. *Gene*, **4**, 121–36.
19. Young, I. G. and Poulis, M. I. (1978) Conjugal transfer of cloning vectors derived from ColEl. *Gene*, **4**, 175–9.
20. Twigg, A. J. and Sherratt, D. (1980) Trans-complementable copy-number mutants of plasmid ColEl. *Nature*, **283**, 216–8.
21. Backman, K., Betlach, M., Boyer, H. W. and Yanofsky, S. (1978) Genetic and physical studies on the replication of ColEl type plasmids. *Cold Spring Harb. Symp. Quant. Biol.*, **43**, 69–76.
22. Meacock, P. A. and Cohen, S. N. (1980) Partitioning of bacterial plasmids during cell division: A cis acting locus that accomplishes stable plasmid inheritance. *Cell*, **20**, 529–42.
23. Jones, I. M., Primrose, S. B., Robinson, A. and Ellwood, D. C. (1980) Maintenance of some ColEl plasmids in chemostat culture. *Mol. Gen. Genet.*, **180**, 579–84.
24. Vieira, J. and Messing, J. (1982) The pUC plasmids, an Ml3mp7 derived for insertion mutagenesis and sequencing with synthetic universal primers. *Gene*, **19**, 259–68.
25. Gray, M., Colot, H. V., Guarente, L. and Rosbash, M. (1982) Open reading frame cloning: identification, cloning, and expression of open reading frame DNA. *Proc. Natn Acad. Sci. USA*, **79**, 6598–602.
26. Zabeau, M. and Stanley, K. K. (1982) Enhanced expression of cro-β-galactosidase fusion proteins under the control of the p_R promoter of bacteriophage λ. *EMBO J.*, **1**, 1217–24.
27. Weinstock, G. M., ap Rhys, C., Berman, M. L. *et al.* (1983) Open reading

frame expression vectors: a general method for antigen production in *Escherichia coli* using protein fusions to β-galactosidase. *Proc. Natn Acad. Sci. USA*, **80**, 4432–6.
28. Koenen, M., Ruther, U. and Muller-Hill, B. (1982) Immunoenzymatic detection of expressed gene fragments cloned in the *lacZ* gene of *Escherichia coli. EMBO J.*, **1**, 509–12.
29. Ruther, U., Koenen, M., Otto, K. and Muller-Hill, B. (1980) pUR222 a vector for cloning and rapid chemical sequencing of DNA. *Nucl. Acids Res.*, **9**, 4087–109.
30. Ruther, U. and Muller-Hill, B. (1983) Easy identification of cDNA clones. *EMBO J.*, **2**, 1791–4.
31. Stanley, K. K. (1983) Solubilization and immune detection of β-galactosidase hybrid proteins carrying foreign antigenic determinants. *Nucl. Acids Res.*, **11**, 4077–92.
32. Jacob, A. E., Cresswell, J. M., Hedges, R. W., Coetzec, J. N. and Beringer, J. E. (1976) Properties of plasmid constructed by the *in vitro* insertion of DNA from *Rhizobium leguminosarum* or *Proteus miribilis* into RP4. *Mol. Gen. Genet.*, **147**, 315–23.
33. Kohn, M., Kolter, R., Thomas, C., Figurski, D. *et al.* (1979) in *Methods in Enzymology*, Vol. 68 (ed. R. Wu), Academic Press, New York.
34. Ditta, G., Stanfield, S., Corbin, D. and Helinski, D. R. (1980) Broad host range DNA cloning system for Gram negative bacteria: construction of a gene bank of *Rhizobium meliloti. Proc. Natn Acad. Sci. USA*, **77**, 7347–51.
35. Bagdasarian, M., Lurz, R., Puckert, B. *et al.* (1981) Specific purpose plasmid cloning vectors. Broad host range, high copy number, RSV1010-derived vectors, and a host vector system for gene cloning in *Pseudomonas. Gene*, **16**, 237–47.
36. Franklin, F. C. H., Bagdasarian, M., Bagdasarian, M. M. and Timmis, K. N. (1981) Molecular and functional analysis of the *Tol* plasmid pWWO from *Pseudomonas putida* and cloning of genes for the entire regulated aromatic ring *meta* cleavage pathway. *Proc. Natn Acad. Sci. USA*, **78**, 7458–62.
37. Worsey, M. J., Franklin, F. C. M. and Williams, P. A. (1978) Regulation of the degradative pathway enzymes coded for by the *Tol* plasmid (pWWO) from *Pseudomonas putida* mt-2′. *J. Bact.*, **134**, 757–64.
38. Hashimoto-Gotoh, T., Franklin, F. C. H., Nordheim, A. and Timmis, K. N. (1981) Specific purpose plasmid cloning vectors. Low copy number, temperature sensitive, mobilisation defective pSC101 derived containment vectors. *Gene*, **16**, 227–35.
39. Bayley, S. A., Duggleby, C. J., Worsey, M. J. *et al.* (1977) Two modes of loss of the *Tol* function from *Pseudomonas putida* mt92. *Mol. Gen. Genet.*, **154**, 203–4.
40. Pickup, R. W., Lewis, R. J. and Williams, P. A. (1983) *Pseudomonas* species mt14, a soil isolate which contains two large catabolic plasmid, one a *Tol* plasmid and one coding for phenylacetate catabolism and mercury resistance. *J. Gen. Microbiol.*, **129**, 153–8.
41. Chatterjee, D. K., Kellogg, S. T., Hamada, S. and Chakrabarty, A. M. (1981) Plasmid specifying total degradation of 3-chlorobenzoate by a modified orthopathway. *J. Bact.*, **146**, 639–46.
42. Don, R. H. and Pemberton, J. M. (1981) Properties of six pesticide

degradation plasmids isolated from *Alcaligenes paradoxus* and *Alcaligenes eutrophus*. *J. Bact.*, **145**, 681–6.

43. Fisher, P. R., Appleton, J. and Pemberton, J. M. (1978) Isolation and characterisation of the pesticide degrading plasmid pJP1 from *Alcaligenes paradoxus*. *J. Bact.*, **135**, 798–804.
44. Chatterjee, D. K., Kellogg, S. T., Watkins, D. R. and Chakrabarty, A. M. (1981) In *Molecular Biology, Pathogenicity and Ecology of Bacterial Plasmids* (eds. S. Levy, R. Clowes and E. Koenig), Plenum Press, New York, London, pp. 519–28.
45. Kellogg, S. T., Chatterjee, D. K. and Chakrabarty, A. M. (1981) Plasmid assisted molecular breeding new technique for enhanced biodegradation of persistant toxic chemicals. *Science*, **214**, 1133–5.
46. Ehrlich, S. D. (1977) Replication and expression of plasmids from *Staphylococcus aureus* in *Bacillus subtilis*. *Proc. Natn Acad. Sci. USA*, **74**, 1680–2.
47. De Vos, W. M., Venema, G., Ganosi, U. and Trautner, T. A. (1981) Plasmid transformation in *Bacillus subtilis*: fate of plasmid DNA. *Mol. Gen. Genet.*, **181**, 424.
48. Gryczan, T. J., Contente, S. and Dubnau, D. (1980) Molecular cloning of heterologous chromosomal DNA by recombination between a plasmid vector and a homologous resident plasmid in *Bacillus subtilis*. *Mol. Gen. Genet.*, **177**, 459–67.
49. Chang, A. C. Y. and Cohen, S. N. (1979) High frequency transformation of *Bacillus subtilis* protoplasts by plasmid DNA. *Mol. Gen. Genet.*, **168**, 111–5.
50. Dubnau, D., Gryczun, T., Contente, S. and Shivakumar, A. G. (1980) in *Genetic Engineering*, Vol. 2 (eds. J. K. Setlow and A. Hollaender), Plenum Press, New York, p. 115.
51. Bingham, A. H. A., Bruton, C. J. and Atkinson, T. (1980) Characterisation of *Bacillus stearothermophilus* plasmid pAB124 and construction of deletion variants. *J. Gen. Microbiol.*, **119**, 109–115.
52. Kreft, J., Bernard, K. and Goebel, W. (1978) Recombinant plasmids capable of replication in *B. subtilis* and *E. coli*. *Mol. Gen. Genet.*, **162**, 59–67.
53. Chang, A. C. Y. and Cohen, S. N. (1974) Genome construction between bacterial species *in vitro*: replication and expression of *Staphylococcus* plasmid genes in *Escherichia coli*. *Proc. Natn Acad. Sci. USA*, **71**, 1030–4.
54. Brammar, W. J., Muir, S. and McMorris, H. (1980) Molecular cloning of the gene for the β-lactamase of *Bacillus licheniformis* and its expression in *Escherichia coli*. *Mol. Gen. Genet.*, **178**, 217–24.
55. Rodico, M. R. and Chater, K. F. (1982) Small DNA free liposomes stimulate transfection of *Streptomyces* protoplasts. *J. Bact.*, **151**, 1078–85.
56. Bibb, M. J., Schothel, J. L. and Cohen, S. N. (1980) A DNA cloning system for interspecies gene transfer in antibiotic producing *Streptomyces*. *Nature*, **284**, 526–31.
57. Snarez, J. E. and Chater, K. F. (1980) DNA cloning in *Streptomyces*: a bifunctional replicon comprising pBR322 inserted into a *Streptomyces* phage. *Nature*, **286**, 527–9.

5 Expression of cloned DNAs in *E. coli* plasmids

In order to achieve efficient expression of a foreign gene in a bacterial cell it is necessary to put that gene under the control of the transcriptional and translational machinery of the host cell. The bacterium *E. coli* is a natural choice for such studies because of our detailed knowledge of the molecular biology of its gene expression. A vast amount of effort has now been put into the development of plasmid vectors to maximize the expression of foreign DNA in *E. coli*. It is outside the scope of this chapter to fully document these efforts, which have been reviewed elsewhere [1]. Instead we will look at essential features of such heterologous expression systems.

A number of factors also have to be taken into account. The eukaryotic gene must be supplied with a prokaryotic promoter. In addition the nucleotide sequence responsible for the binding of mRNA to ribosomes differs in prokaryotes from that found in eukaryotes. The ribosome binding site will also have to be supplied by the cloning vector. Furthermore the genes of *E. coli* do not have intervening sequences and so the bacterium has no means of removing such sequences from primary transcripts. The eukaryotic gene should therefore be either a cDNA copy of mature mRNA or a chemically synthesized gene not containing intervening sequences. Each of these difficulties can be overcome by the careful design of *in vitro* recombinant molecules and by using an appropriate vector.

5.1 Promoters

The genes of eukaryotic organisms differ in a number of fundamental ways from those of *E. coli*. In order for eukaryotic genes to be expressed in *E. coli* they must be modified in such a way that their organization resembles prokaryotic genes. Many of these differences lie in controlling elements upstream of the structural gene. The analysis of promoter sequences upstream from the transcription initiation sites of many genes from *E. coli* has revealed two blocks of conserved sequence. These are the sequences TTGACA and TATAAT, which are found at positions −35 (the '−35 sequence') and −10 (the Pribnow box) from the initiation site respectively. They are involved in the binding of RNA polymerase. The number of nucleotides that separate the blocks is important and the efficiency of the promoter is decreased by mutations that alter the spacing. An analogous sequence TATAA, is found about 30 nucleotides upstream from eukaryotic genes. In both eukaryotic and prokaryotic genes additional 5′ cis acting DNA sequences play a part in controlling gene expression. In order for a eukaryotic gene to be transcribed within a prokaryotic cell, it is necessary to replace the eukaryotic promoter element with a prokaryotic promoter. Most plasmid expression vectors utilize either the promoter of the *lac* or of the *trp* operons from *E. coli* or the β-lactamase promoter from pBR322.

5.1.1 The *lac* promoter

We have previously encountered a number of plasmid vectors which utilize segments of the *lac* operon including the regulatory region. These vectors can in principle be used as expression vectors. Most of these vectors contain enough of the *lacZ* gene to encode the α-peptide of β-galactosidase. Thus, if following the insertion of DNA the reading frame of the *lacZ* gene is maintained then a fusion peptide is produced. This can be detected by intra-allelic complementation of a *lacZ* gene with an N-terminal deletion in the host bacterium. Other examples of the production of fusion proteins will be encountered in Section 5.2. Alternative vectors have been also developed which contain the minimum controlling elements from the *lac* operon.

The *lac* operon is negatively regulated by the *lac* repressor (the *lacI* gene product), and can be derepressed by addition of an inducer, such as isopropyl β-D-thiogalactoside (IPTG). It is also positively regulated by the catabolite activator protein (CAP) complexed with cyclic AMP [2]. The control regions of the *lac* operon which interact with the repressor, the catabolite activator protein and with RNA polymerase are all contained on a 203 base pair *Hae*III fragment (see Fig. 5.1). This 203 base pair *Hae*III fragment has been isolated from *lac* transducing phages and incorporated into plasmids to provide a promoter for the expression of foreign genes. The fragment contains the operator, promoter and first eight codons of β-galactosidase. It was originally cloned by blunt-end ligation into *Eco*RI cleaved pBR322, in which the *Eco*RI termini had been repaired using DNA polymerase. This procedure regenerates *Eco*RI sites (see Fig. 5.1), thereby allowing the fragment to be easily moved from one vector to another [3, 4]. Alternative fragments have also been used in which the promoter is insensitive to catabolite repression [5]. This is achieved by either using the internal 95 base pair *Alu*I fragment (Fig. 5.1) which does not contain the intact sequence required for CAP binding, or by using the 203 base pair *Hae*III fragment containing the L48 mutation. This mutation is a G to A transition within the 14 nucleotide sequence that interacts with CAP. The *Hae*III fragment also contains a second mutation, uv5, which leads to more efficient transcription from the promoter. This is a G to A transition adjacent to a T to A transversion in the RNA polymerase binding site. The promoter fragment also contains the 5′ coding sequence of the *lac*Z gene and so recombinant molecules generally specify a fusion protein, having the *N*-terminal sequence of β-galactosidase (see also below). Charnay *et al.* [6] have constructed a set of vectors which allow fusion of the cloned genes in each of the three translational phases. They took the *Eco*RI fragment carrying the uv5 promoter, digested away the cohesive termini with S1 nuclease and ligated it to an octanucleotide linker containing an *Eco*RI site. Cleavage with *Eco*RI generates a fragment two base pairs longer than the original. The procedure was repeated in order to extend the fragment by another two bases (see Fig. 5.1). These two and four base extensions have the effect of changing the translation frame to give all possible combinations when a given *Eco*RI fragment is inserted into the three vectors. Additional promoter fragments have been derived in which the β-galactosidase codons have been deleted [6]. These derivatives are suitable for the expression of non-fusion proteins.

5.1.2 The *trp* promoter

A similar series of promoter elements from the *trp* operon have also been incorporated into expression vectors. The *trp* operon encodes five enzymes for the biosynthesis of tryptophan. It is under the negative control of a complex formed between the *trp* repressor and tryptophan, so that when the amino-acid is in abundant supply the genes responsible for its biosynthesis are repressed. In addition there is a second level of negative control imposed by the attenuator site. This lies within the transcribed leader sequence that precedes the initiation codon of the *trp* E gene. The attenuator contains a rho-independent termination site that is recognized by RNA polymerase in the presence of tryptophan. The region contains a ribosome binding site followed by a coding region of 13 codons with two

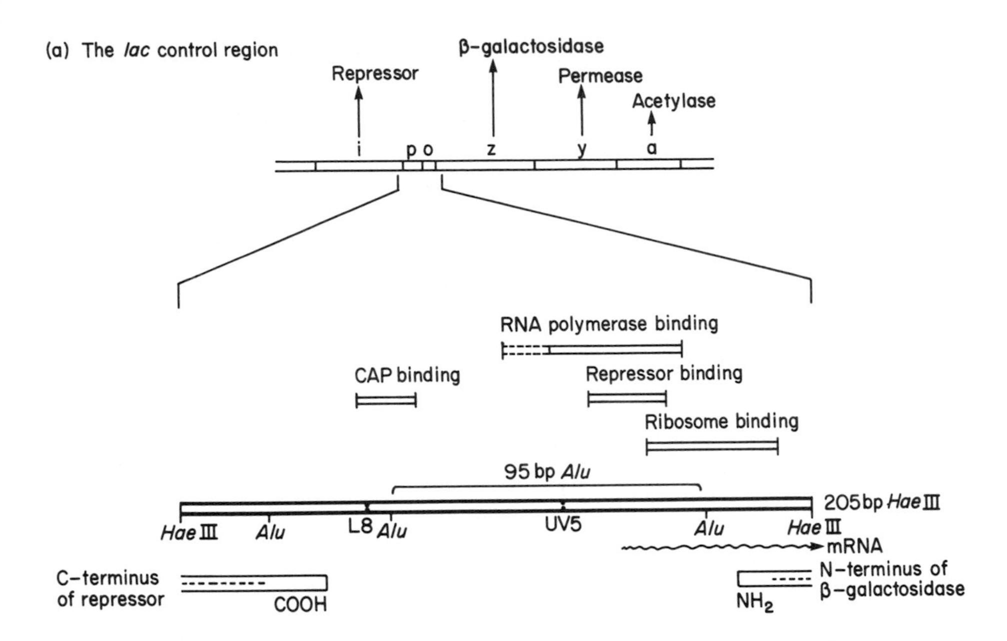

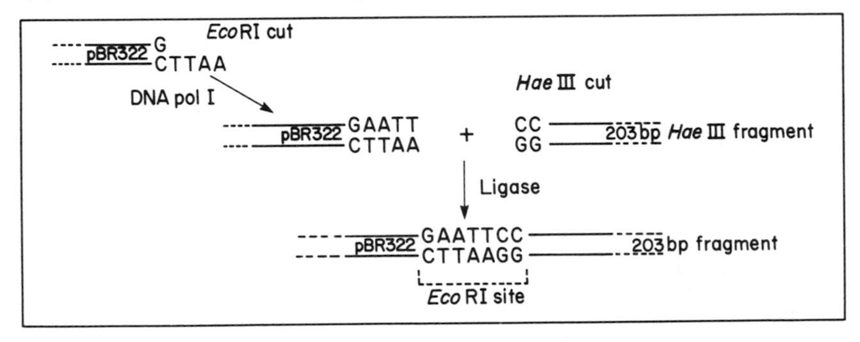

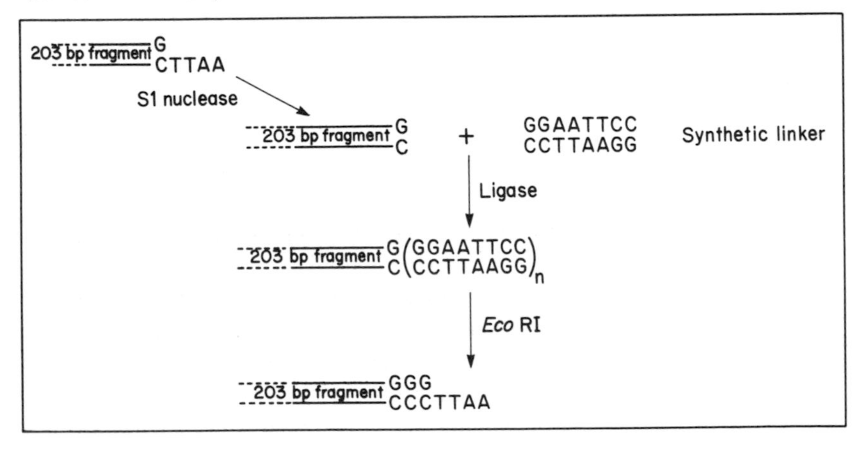

Figure 5.1 The promoter from the *lac* operon.

codons for tryptophan in succession. When levels of tryptophan are limiting, ribosomes stall at these two codons, thereby preventing the recognition of the transcription terminator by RNA polymerase [7]. The stimulus for the development of expression vectors using the *trp* promoter was the early observation that when the *trp* operon was cloned onto a ColEl type vector it directed the synthesis of 25% of the cell protein following induction with 3-indolylacrylic acid (IAA). One of the first *trp* expression vectors was constructed by cloning a *Hin*dIII fragment containing the *trp* promoter, operator, leader, attenuator, *E* gene and part of the *D* gene into pBR322. The recombinant has two *Hin*dIII sites and one of these was subsequently deleted following partial *Hin*dIII digestion followed by treatment with exonuclease III and S1 nuclease [8]. This plasmid, p*trp*ED5.1 (see Fig. 5.2), has been used to express a fused gene encoding the *N*-terminus of the *trp*D protein linked to human growth hormone. In this case about 3% of total cell protein was synthesized as the hybrid protein [9]. A *Hin*fI fragment from p*trp*ED5.1 has been cloned using *Hin*dIII linkers into the *Hin*dIII site of pBR322 (Fig. 5.2). This recombinant, pWT111 contains the *trp* regulatory sequences and the first seven codons of the *trp* E gene. Derivatives of pWT111 have been constructed which allow the expression of a fusion protein in the other two translation frames [10]. The approach was similar to that described above [6] for the phased β-galactosidase fusions. These vectors have been successfully used for the expression of eukaryotic genes. The haemagglutinin gene of fowl plague virus has, for example, been cloned in such a vector [11]. The production of the haemagglutinin was however at lower levels than expected. This has been postulated to be due either to a toxic effect of the eukaryotic sequence responsible for the export of the protein from cells, or to proteolytic cleavage and degradation of the protein in bacterial cells.

5.1.3 The '*tac*' promoter

The '*tac*' promoter is a hybrid promoter that has now been constructed between elements of the *trp* and *lac* promoters [12]. The first of these, *tac*I, contains the '-35 sequence' from the *trp* promoter and the Pribnow box from the *lac* promoter. The second, *tac*II, also has the *trp* '-35 sequence' but the Pribnow box is a hybrid between the *trp* and *lac* promoters. *ptacI* and *ptacII* respectively direct transcription eleven and seven times more efficiently than the derepressed *lac*uv5 promoter and three and two times more efficiently than the *trp* promoter. They are both subject to repression by the *lac* repressor and derepression by IPTG. The hybrid promoter more closely matches the consensus '-35 sequence' and Pribnow box than either of the parental promoters. The spacing of the boxes is also

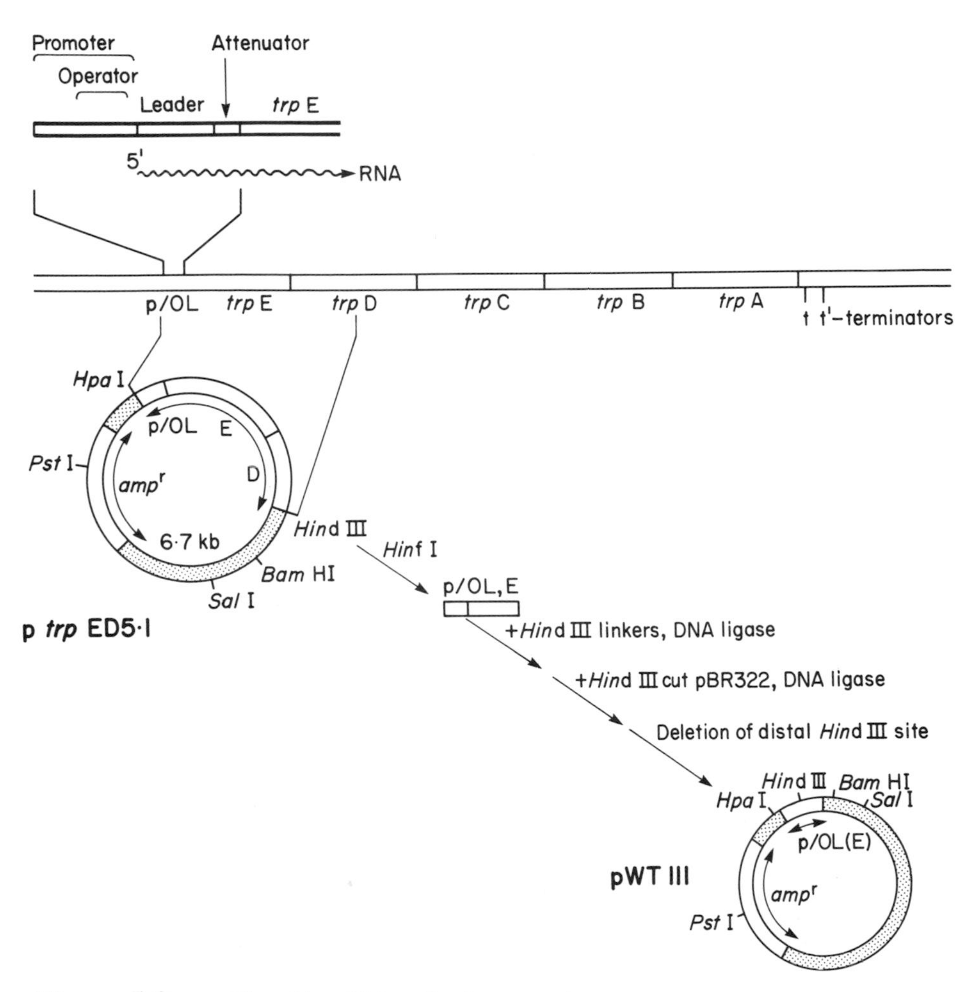

Figure 5.2 Plasmids utilizing the *trp* promoter.

closer (Fig. 5.3). Each of these factors could account for the increased efficiency of the hybrid promoters.

5.2 Fusion proteins – a means to increase the stability of foreign peptides in *E. coli*

5.2.1 Fusion proteins that can be cleaved by CNBr

Somatostatin was the first polypeptide to be expressed in *E. coli* as part of a fusion peptide [4]. The tetradecapeptide hormone, somatostatin, has the physiological role of inhibiting the secretion of growth hormone, glucagon and insulin. The gene was chemically synthesized, using the triester method (see Chapter 2), as eight

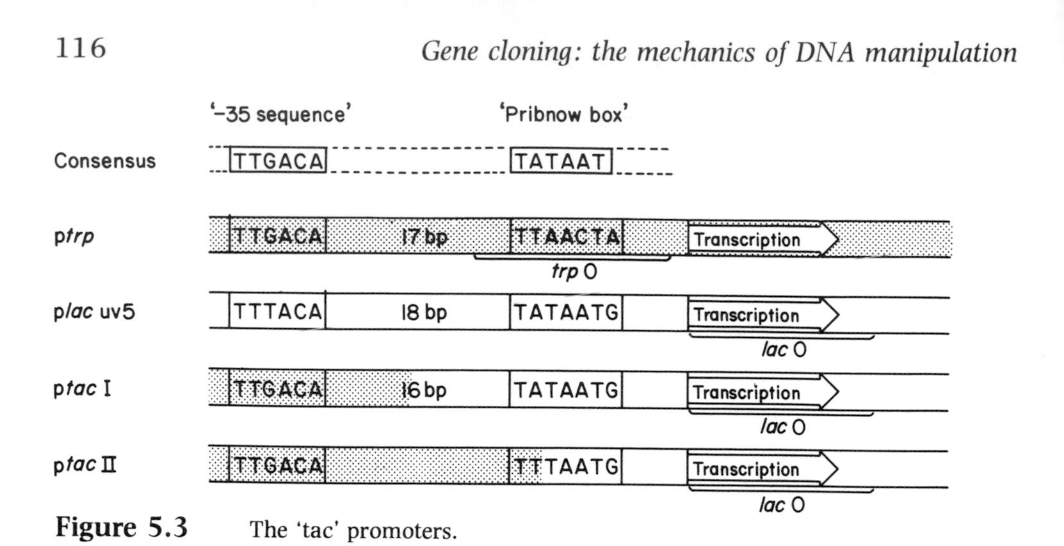

Figure 5.3 The 'tac' promoters.

single-stranded DNA segments which anneal in an overlapping manner to give a double-stranded DNA segment having single-stranded projections at each end corresponding to the cohesive ends produced by *Eco*RI and *Bam*HI. The codons were chosen so as to eliminate undesirable base pairing of the eight synthetic segments and at the same time to include those codons which are used preferentially in *E. coli*. The synthetic gene was terminated by two nonsense codons and preceded by a methionine codon. Somatostatin does not contain any internal methionine residues and so the fused polypeptide can be cleaved by cyanogen bromide to produce functional hormone (Fig. 5.4.). Two plasmids were constructed with the synthetic gene inserted into the *E. coli* β-galactosidase gene at different sites. In the first of these, pSoml (Fig. 5.4), the synthetic gene was linked to the 203 nucleotide *Hae*III fragment carrying the *lac* promoter modified to have *Eco*RI termini. The chemically synthesized gene was inserted downstream from the *lac* promoter in such a way that the gene fusion should have specified a polypeptide in which the first seven amino-acids of β-galactosidase were fused to somatostatin. It was not possible, however, to detect somatostatin in bacteria transformed by this recombinant molecule. This was thought to be due to proteolytic degradation of the peptide. An alternative plasmid, pSomII.3, was therefore constructed in which the synthetic gene was inserted at an *Eco*RI site near the *C*-terminus of β-galactosidase. This directs the synthesis of a fused protein stabilized from proteolytic degradation by the β-galactosidase moiety. Somatostatin can be prepared from extracts of cells carrying pSomII.3 by treatment with CNBr.

This approach should be generally applicable to proteins which do not contain internal methionines. The analogous route has also been followed for human insulin [13]. Insulin is naturally synthesized as a

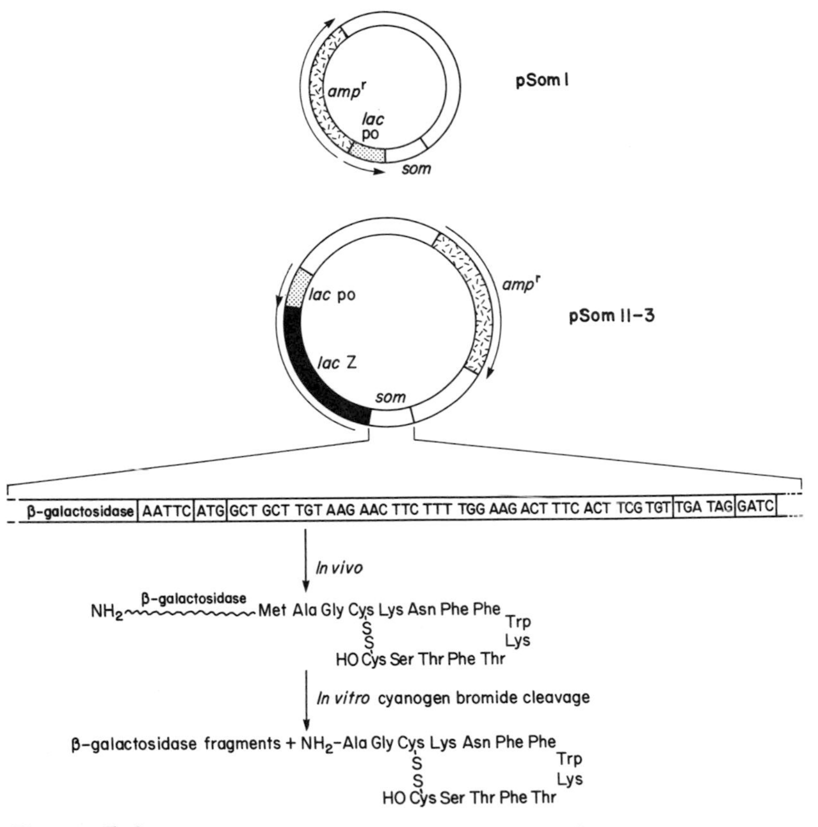

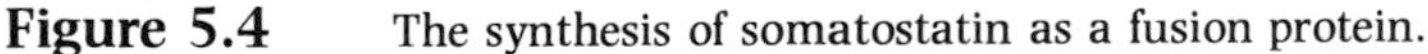

Figure 5.4 The synthesis of somatostatin as a fusion protein.

single peptide, pre-proinsulin. It has an *N*-terminal signal peptide responsible for the transport of the molecule across the membranes of storage vesicles. The signal peptide is removed during this process to generate proinsulin. Proinsulin is converted into insulin inside the vesicles by proteolytic cleavage. This leaves the mature A and B chains, from the *N*- and C-termini respectively, held together by disulphide bridges. In order to make insulin in bacteria, two synthetic genes have been constructed which encode the 21 amino acid A chain and the 30 amino acid B chain. These were each chemically synthesized with *Bam*HI and *Eco*RI ends and cloned into pBR322. The same fragment of the *lac* operon was used to promote expression as with the successful somatostatin experiment described above. The hybrid proteins produced were insoluble and accounted for about 20% of the cell protein. They could, however, be dissolved by successive treatment with guanidinium hydrochloride and formic acid and then cleaved at methionine residues by treatment with cyanogen bromide. In order to reconstitute native insulin, the peptides then must be

S-sulphonated using sodium dithionate and sodium sulphite. Insulin has now been produced by a scaled-up version of this process and shown to be active when injected into humans [14].

5.2.2 Fusion proteins with internal methionine residues

An alternative strategy is needed for genes whose proteins have internal methionine residues. One possibility has been explored with the gene for β-endorphin, although it seems unlikely that this approach could be used economically on a large scale [15]. β-endorphin is a 30 amino-acid neuropeptide with opiate activity. It is proteolytically cleaved from a precursor peptide which also contains the hormones α-melanotropin (α-MSH), corticotropin (ACTH), β-lipotropin (β-LPH), and β-melanotropin (β-MSH). The β-endorphin sequence is at the C-terminus of the precursor peptide. The DNA which encodes it can be cleaved from cDNA encoding the precursor using *Hin*dIII and *Hpa*II. Unfortunately, the *Hin*dIII cleavage removes the C-terminal glutamine codon. In order to regenerate this codon and place the gene in correct translational phase when fused to β-galactosidase, the cohesive termini were partially filled in with dATP and dCTP using reverse transcriptase (see Fig. 5.5). The ends were trimmed with S1 nuclease and the fragment blunt-end ligated to synthetic linkers containing an *Eco*RI site. This creates a stop codon following the C-terminal glutamine (Fig. 5.5). The *Eco*RI fragment was subsequently cloned into the 3′ region of the *lac*Z gene. Bacteria transformed by such plasmids produce an insoluble fusion protein between β-galactosidase and β-endorphin. β-endorphin can be liberated from the hybrid protein by trypsin which cleaves only at the arginine residue preceding the hormone. This necessitates first protecting the internal lysines from trypsinization by the reversible process of citraconylation. Two basic amino-acids are commonly found at the junctions between hormones within precursor proteins and so the procedure may in future find general application.

Trp expression vectors have also been used to direct the synthesis of insoluble hybrid protein. Immunogenic capsid protein of foot and mouth disease virus, for example, has been expressed in a plasmid containing the *N*-terminus of the *trp* leader peptide fused to the *C*-terminus of the *trp*E gene in which there is an *Eco*RI site [16]. A 609 nucleotide fragment of the capsid protein gene was cloned into this site using linkers to maintain the translation frame. The resulting plasmid specifies an insoluble 44 000 dalton fusion protein which has been used directly as a immunogen in pigs and cattle.

Antigenic determinants of Hepatitis B virus have also been expressed in *E. coli* as fusion proteins. The production of these

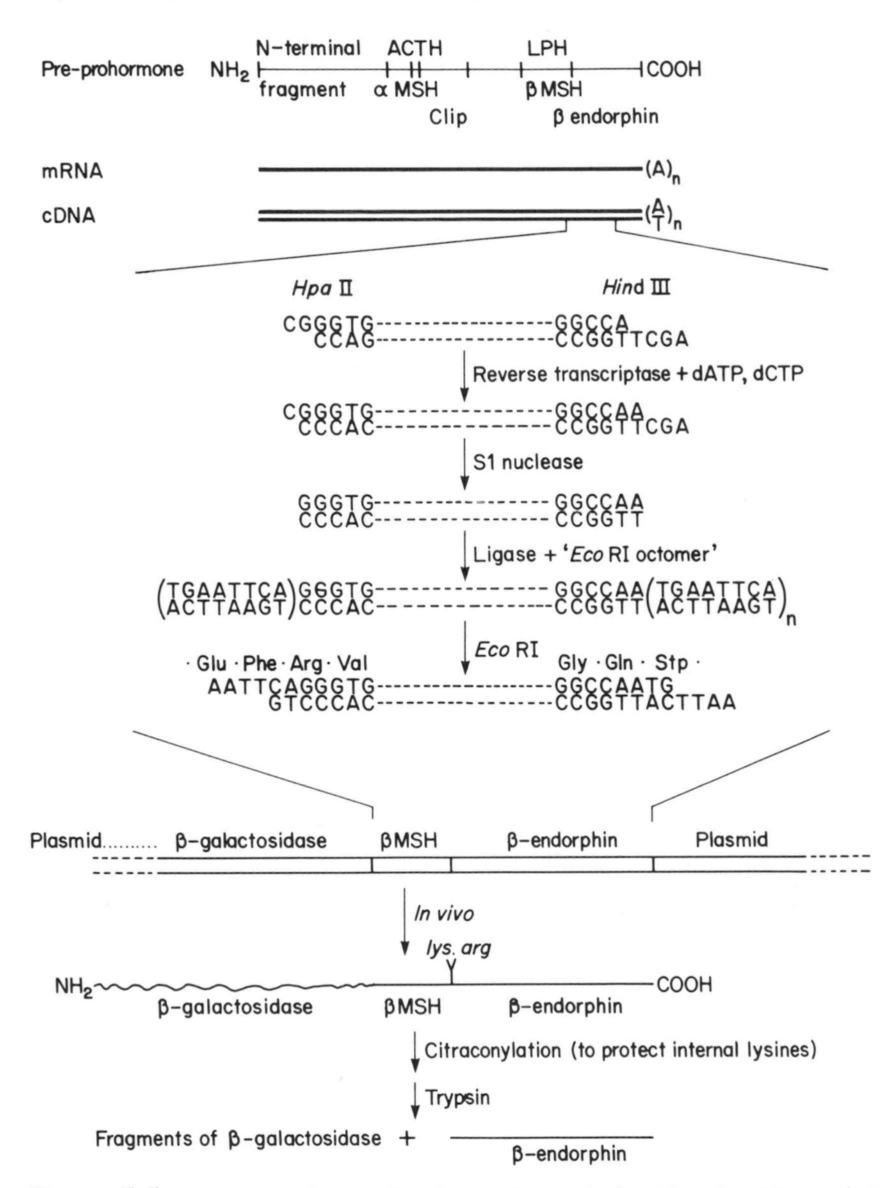

Figure 5.5 A recombinant that directs the synthesis of β-endorphin as a fusion peptide.

antigens is needed in order to facilitate vaccine production. The virus is widespread in man and produces several types of chronic liver disorders. The limited amount of virus available from infected patients and the inability to grow the virus in cultured cells has seriously hindered its molecular characterization and the development of vaccines. The viral DNA is a double-stranded circular molecule of approximately 3 kb and has a large single-stranded gap which must be repaired with an endogenous viral polymerase before the molecule can be digested with restriction endonucleases for molecular cloning. Three viral proteins have been recognized antigenically: a viral surface antigen, an inner viral core antigen and a third antigen, the e-antigen, which is detected in the blood of some infected individuals. The complete viral genome has been cloned and sequenced. Limited knowledge of the amino-acid sequence at the termini of the surface and core antigens has enabled their genes to be localized within the complete nucleotide sequence of the viral DNA. The gene for the core antigen has been cloned and expressed from the β-lactamase promoter [17], and the *trp* and *lac* [18, 19] promoters. Small amounts of the surface antigen have been produced when this gene has been fused to either the 5′ region of the β-lactamase gene [20] or the β-galactosidase gene [18]. These antigens appear, however, to be synthesized more efficiently in yeast or mammalian systems.

5.3 Secretion of proteins from *E. coli*

A number of proteins are naturally secreted from the cytoplasm of *E. coli* into the periplasmic space. A foreign protein secreted in such a way might be more easily purified and more easily detected by the direct immunological screening technique. In Section 4.1.6 we encountered a vector designed to detect open reading frames which directs the synthesis of fusion proteins having the *N*-terminus of an outer membrane protein. This *N*-terminal region contains a 'signal peptide' which enables such proteins to cross the cell membrane. Many proteins that determine resistance to antibiotics are exported by means of *N*-terminal signal sequences. β-lactamase, for example, is a periplasmic protein synthesized with a 23 amino-acid leader or signal peptide, which directs the protein across the cell membrane and which is removed during this process. A number of laboratories have cloned cDNAs into the *Pst*I site within the β-lactamase gene. The approach has been to tail the cDNA with dC residues and the vector with dG residues in order to reconstitute the *Pst*I site. Villa-Kamaroff and colleagues have cloned insulin cDNA in this way [21]. Some of their recombinants encode hybrid polypeptides of insulin with the *N*-terminus of β-lactamase. The two domains are connected by six glycine residues encoded by the dG:dC connector region. This fusion protein is exported from the cytoplasm into the periplasmic space because it has the signal peptide of β-lactamase. This facilitates the detection of colonies of cells carrying recombinant plasmids by the solid phase radioimmune assay.

Seeberg and coworkers [22] have taken advantage of the fortuitous *Pst*I site in codon 24 of rat pre-growth hormone cDNA.

This is the region of the gene that encodes the eukaryotic signal sequences responsible for directing the transport of the newly synthesized hormone out of pituitary cells. The rat growth hormone gene can be fused to the β-lactamase gene through their *Pst*I sites without losing the translational reading frame. The hybrid gene then encodes a fusion peptide with 181 amino-acids from the *N*-terminus of β-lactamase and 214 amino-acids of rat pre-growth hormone as its *C*-terminus.

The transport of proteins through the cell membrane in both prokaryotic and eukaryotic cells is dependent upon the presence of several hydrophobic amino-acids in the *N*-terminal region signal peptide. The sequences necessary for the transport process have been investigated by Talmadge and colleagues [23] who generated a series of deletions around the *Pst*I site in the β-lactamase gene using *Bal*31 nuclease. The deletions extended up to or within the signal peptide coding sequences. *Pst*I sites were regenerated in these plasmids using *Pst*I linkers. A series of recombinants were then constructed between these plasmids and the genes for rat preproinsulin, in which the signal peptide encoding sequences are present, or proinsulin, in which the signal sequences are absent. From the amounts of insulin transported into the periplasmic space it was concluded that either the prokaryotic or the eukaryotic signal sequence could direct the transport of the protein. The functional capability of the eukaryotic signal sequence was further demonstrated in a recombinant in which the rat preproinsulin was fused to the *E. coli lacZ* gene to direct the synthesis of a hybrid protein in which the *N*-terminal eight amino-acids were those of β-galactosidase [24]. This hybrid protein does not have a prokaryotic signal peptide, and the eukaryotic signal peptide is not precisely at the *N*-terminus. It is nevertheless transported into the periplasmic space. It seems that proinsulin in the periplasmic space is less rapidly degraded than proinsulin which remains in the cytoplasm [25]. This may well be an important consideration in terms of the commercial application of these techniques.

5.4 Factors affecting the translation of hybrid genes

Prokaryotic mRNAs have a sequence upstream from the initiation codon which is believed to play a role in the attachment of the 30 S ribosomal subunit to the mRNA [26]. It is 3–9 bases long and 3–12 bases upstream from the initiation codon. This sequence is known as the Shine–Dalgano sequence and is complementary to a sequence near the 3′ terminus of 16S ribosomal RNA. Translation is further influenced by the degree of secondary structure of the mRNA at its 5′ end.

In order to achieve efficient expression of a eukaryotic gene within *E. coli* it is necessary to provide that gene with a Shine–Dalgano sequence. The sequence is usually contained within the expression vector downstream from the promoter. It seems that in some cases, however, the necessary requisites for efficient translation have been fortuitously created in the dG:dC connectors used when cloning cDNA. Transcripts of cDNA cloned into the *Pst*I site of pBR322 are not always translated to give fused polypeptides (as with the insulin gene). Recombinants have been described in which translation is correctly initiated at the AUG codon in the cDNA. The biological expression of a DNA sequence encoding the mouse dihydrofolate reductase (DHFR) is one such example. DHFR catalyses the conversion of dihydrofolic acid to tetrahydrofolic acid. When mammalian cell lines are selected for their ability to grow in the presence of methotrexate they are found to overproduce this enzyme. This is a consequence of the amplification of the DHFR gene by greater than 100-fold in these cells. The mRNA for DHFR is readily purified from these cells. cDNA synthesized from such a preparation of mRNA has been cloned by the dG:dC tailing technique into the *Pst*I site of pBR322 [27]. Bacterial cells producing the functional mammalian enzyme can be selected using the drug trimethoprim. This binds to the mammalian enzyme with a lower affinity than it does to the bacterial enzyme. *E. coli* cells carrying functional recombinant plasmids will therefore grow in a higher concentration of the drug than their counterparts having only the *E. coli* enzyme. Independent clones containing recombinant plasmids were found to show differing degrees of resistance to trimethoprim. In the case of one clone which showed resistance to a high concentration of the drug, the plasmid vector was shown to be interrupted at its *Pst*I side by 11 dG residues followed by an AUG codon and then the codon for the first amino acid of DHFR, which is no longer the same translational reading frame as the β-lactamase gene. The polypeptide produced in these bacteria was identical to the mammalian enzyme in a number of antigenic tests, in its sensitivity to methotrexate and trimethoprim and in its electrophoretic mobility. It seems therefore that in this recombinant molecule, the poly dG sequence serves as a Shine–Dalgano sequence [26]. In other recombinant molecules with differing numbers of residues between the *N*-terminus of the gene and the dG:dC connector sequences the gene was not as efficiently expressed.

The biosynthesis of herpes simplex virus thymidine kinase [28], hepatitis B core antigen [17] and leukocyte interferon [24] have also been shown to be independent of the translational reading frame when cloned into the *Pst*I site of pBR322. In both these cases, Shine–Dalgano sequences can be recognized upstream from the eukaryotic initiation codon, suggesting that translation is being initiated independently of β-lactamase.

In the above cases the correct initiation of translation at the AUG codon of the eukaryotic gene was somewhat fortuitous. Specific constructs have however been designed which place the initiator codon immediately downstream of the Shine–Dalgano sequence, associated with the promoter element. This has led to several studies on the efficiency of translation when the distance between the Shine–Dalgano sequence and the AUG is varied. Some of the first experiments of this kind were carried out with cloned DNAs in which the bacteriophage λ genes *cI* and *cro* were placed under the transcriptional control of the *lac* promoter [3, 20]. The studies with the *cro* gene are most impressive, some of the recombinants producing as much as 200 000 *cro* protein molecules per cell, the equivalent of 1.6% of the total cell protein. The amount of *cro* protein produced depended upon the distance between the *lac* promoter and Shine–Dalgano sequence and the *cro* gene in a particular construct. The correlation is not simply dependent upon length, however, since very small differences in length of two or three nucleotides between the Shine–Dalgano sequence and AUG codon could affect the efficiency of translation by an order of magnitude. It was concluded that the nucleotide sequence itself at the 5′ end of the message was important since this dictates the secondary structure of the message. If the Shine–Dalgano sequence or AUG codon participates in base pairing within the mRNA this leads to considerable loss in the efficiency with which that mRNA is translated.

Several groups have used analogous approaches with cloned mammalian genes. Goeddel and his coworkers [29], designed a recombinant plasmid in which translation was initiated at the natural Met codon of the human growth hormone gene. Since the mature hormone has 191 amino acids, the chemical synthesis of its gene would be too time-consuming. A gene was therefore constructed, between a restriction fragment of cDNA synthesized from pre-hormone mRNA and a chemically synthesized DNA segment corresponding to the *N*-terminal region. Double-stranded cDNA was made against total polyadenylated RNA from human pituitaries and cleaved with *Hae*III. A 550 base pair *Hae*III fragment was gel purified, tailed with dC residues and cloned into *Pst* cleaved pBR322 tailed with dG residues (pHGH31). This procedure regenerates the *Hae*III sites (GGCC) on the ends of the cDNA and the *Pst*I site (CTGCAG) of pBR322. The *Hae*III site occurs at amino-acid residues 23 and 24 and so it was necessary to chemically synthesize double-stranded DNA which would encode the sequence up to these residues. This segment was synthesized with *Eco*RI and *Hin*dIII termini so that it could be cloned between these sites in pBR322 (pHGH3). The third element of the construct was a 285 base pair fragment containing two *lac* promoter elements, cloned into pBR322. The three regions were then combined as outlined in Fig. 5.6. The *lac* promoter segment contains a ribosome binding sequence AGGA which occurs 11 base pairs in front of the initiation codon of the human growth hormone gene in the final recombinant, pHGH107. In the *lac*Z gene this sequence

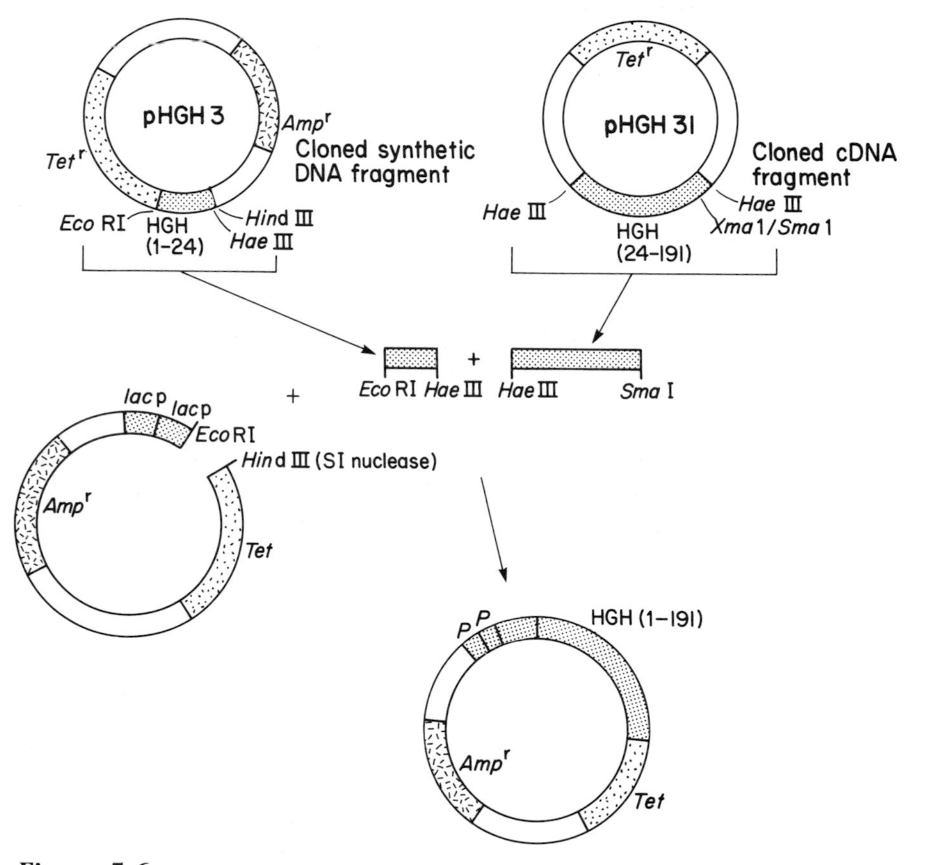

Figure 5.6 Construction of a plasmid that directs the synthesis of human growth hormone.

occurs seven base pairs in front of the initiation codon. A derivative of pHGH107 was therefore constructed in which four nucleotides from this region were deleted. This was achieved by cleaving the plasmid with *Eco*RI, removing the single-stranded tails with S1 nuclease and religating the blunt ends with T4 ligase. Unexpectedly it was found that human growth hormone was more efficiently produced in the plasmid in which there were 11 residues between the putative ribosome binding site and the methionine codon. The experiment again illustrates that the effectiveness whereby the transcription products of the *in vitro* recombinant gene are translated does depend in a somewhat unpredictable manner upon the sequence which immediately precedes the initiator codon. The human growth hormone is produced as a soluble protein which can be readily purified from *E. coli* and has been shown to have biological activity in man [30].

A similar procedure has been followed by one of the many groups to have cloned human interferon genes into expression vectors [33]. Interferons are a family of proteins induced by viral infections and which confer viral resistance upon their target cells. They are classified into two major groups, leukocyte interferon (LeIF) and fibroblast interferon (FIF) according to their cells of origin. Each group represents a family of related genes. Goeddel *et al.* [31] joined a synthetic oligonucleotide onto a restriction fragment from the 5′ end of the LeIF-A gene, so as to recreate the codon for the *N*-terminal amino-acid linked to the initiator codon and *Eco*RI site. This fragment was then annealed to fragments containing the rest of the coding sequence and inserted into a plasmid containing the *trp* promoter, leader and ribosome binding site. This construct directs the efficient synthesis of a soluble protein with all the properties of leukocyte interferon, including the ability to protect monkeys from infection by encephalomyocarditis virus. Similar experiments have been carried out to direct the expression of the FIF gene from both *trp* and *lac* promoters [32]. Once again the expression of both LeIF and FIF has been optimized by varying the spacing between the *trp* Shine–Dalgano sequence and the initiator codon. In both cases the optimal spacing was nine nucleotides.

The different members of the interferon gene family have different antiviral specificities. Different members of the LeIF gene family have been placed in the expression vectors and shown to direct the synthesis of proteins with the appropriate antiviral properties. This has led to the construction of hybrid genes consisting of the *N*-terminal coding sequences of one type linked to the *C*-terminal coding sequences of another. These hybrid genes direct the synthesis of interferons having biological characteristics which differ from either of the parent molecules [33].

5.5 Post-translational modification

Glycosylation and phosphorylation are common modifications of proteins in eukaryotic cells, and can be important for the biological activity of the protein. These modifications do not occur in prokaryotes. This is an area which is therefore likely to be a rate-limiting step for the expression of functional polypeptides from several eukaryotic genes. In cases where the post-translational modification is of prime importance, it may be better therefore to turn to a eukaryotic expression system.

References

1. Harris, T. J. R. (1983) in *Genetic Engineering 4* (ed. R. Williamson), Academic Press, London, pp. 128–75.
2. Dickson, R. C., Abelson, J., Barnes, W. H. and Reznikoff, W. S. (1975) Genetic regulation, the *lac* control region. *Science*, 187, 27–35.
3. Backman, K. and Ptashne, M. (1978) Maximising gene expression on a plasmid using recombination *in vitro*. *Cell*, **13**, 65–71.
4. Itakura, K., Hirose, T., Crea, R. *et al.* Expression in *Escherichia coli* of a chemically synthesised gene for the hormone somatostatin. *Science*, **198**, 1056–63.
5. Fuller, F. (1982) A family of cloning vectors containing the *lac*UV5 promoter. *Gene*, **19**, 43–54.
6. Charnay, P., Perricaudet, M., Galibert, F. and Tiollais, P. (1978) Bacteriophage λ and plasmid vectors allowing fusion of cloned genes in each of the three translational phases. *Nucl. Acids Res.*, **5**, 4479–94.
7. Yanofsky, C. (1981) Attenuation in the control of expression of bacterial operons. *Nature*, **289**, 751–8.
8. Hallewell, R. A. and Emtage, S. (1980) Plasmid vectors containing the tryptophan operon promoter suitable for efficient regulated expression of foreign genes. *Gene*, **9**, 27–47.
9. Martial, J. A., Hallewell, R. A., Baxter, J. D. and Goodman, H. M. (1979) Human growth hormone: complementary DNA cloning and expression in bacteria. *Science*, **205**, 602–7.
10. Tacon, W., Carey, N. H. and Emtage, S. (1980) The construction and characterisation of plasmid vectors suitable for the expression of all DNA phases under the control of the *E. coli* tryptophan promoter. *Mol. Gen. Genet.*, **177**, 427–38.
11. Emtage, J. S., Tacon, W. C. A., Catlin, G. H. *et al.* (1980) Influenza antigenic determinants are expressed from haemagglutinin genes cloned in *Escherichia coli*. *Nature*, **283**, 171–4.
12. De Boer, H. A., Comstock, L. J. and Vasser, M. (1983) The *tac* promoter: a functional hybrid derived from the *trp* and *lac* promoters. *Proc. Natn Acad. Sci. USA*, **80**, 21–5.
13. Goeddel, D. V., Kleid, D. G., Bolivar, F. *et al.* Expression in *Escherichia coli* of chemically synthesised genes for human insulin. *Proc. Natn Acad. Sci. USA*, **76**, 106–10.
14. Clark, A. J. L., Adeniyi-Jones, R. O., Knight, G. *et al.* (1982) Biosynthetic human insulin in the treatment of diabetes. A double blind cross-over trial in established diabetic patients. *Lancet*, **2**, 354–7.
15. Shine, J., Fettes, I., Lan, N. C. Y., Roberts, J. L. and Baxter, J. D. (1980) Expression of cloned β-endorphin gene sequences by *Escherichia coli*. *Nature*, **285**, 456–61.
16. Kleid, P. G., Yansura, D., Small, B. *et al.* (1981) Cloned viral protein

vaccine for foot and mouth disease: responses in cattle and swine. *Science*, **214**, 1125–9.
17. Burrell, C. J., Mackay, P., Greenaway, P. J., Hofschneider, P. and Murray, K. (1979) Expression in *Escherichia coli* of hepatitis B DNA sequences cloned in plasmid pBR322. *Nature*, **279**, 43–7.
18. Edman, J. C., Hallewell, R. A., Valenzuela, P., Goodman, H. M. and Rutter, W. J. (1981) Synthesis of hepatitis B surface and core antigens in *E. coli*. *Nature*, **291**, 503–6.
19. Stahl, S., Mackay, P., Magazin, M., Bruce, S. A. and Murray, K. (1982) Hepatitis B virus core antigen: synthesis in *Escherichia coli* and application in diagnosis. *Proc. Natn Acad. Sci. USA*, **79**, 1606–10.
20. Mackay, P., Pasek, M., Magazin, M. *et al.* (1981) Production of immunologically active surface antigens of hepatitis B virus by *Escherichia coli*. *Proc. Natn Acad. Sci. USA*, **78**, 4510–4.
21. Villa-Kamaroff, L., Efstratiadis, A., Broome, S. *et al.* A bacterial clone synthesising proinsulin. *Proc. Natn Acad. Sci. USA*, **75**, 3727–31.
22. Seeberg, P. H., Shine, J. H., Martial, J. A. *et al.* (1978) Synthesis of growth hormone by bacteria. *Nature*, **276**, 795–8.
23. Talmadge, K., Kaufman, J. and Gilbert, W. (1980) Bacteria mature preproinsulin to proinsulin. *Proc. Natn Acad. Sci. USA*, **77**, 3988–92.
24. Talmadge, K., Brosias, J. and Gilbert, W. (1981) An internal signal sequence directs secretion and processing of proinsulin in bacteria. *Nature*, **294**, 176–8.
25. Talmadge, K. and Gilbert, W. (1982) Cellular location affects protein stability in *Escherichia coli*. *Proc. Natn Acad. Sci. USA*, **79**, 1830–3.
26. Shine, J. and Dalgano, L. (1975) Determinant of cistron specificity in bacterial ribosomes. *Nature*, **254**, 34–8.
27. Chang, A. C. Y., Nunberg, J. H., Kaufman, R. J. *et al.* (1978) Phenotypic expression in *E. coli* of a DNA sequence coding for mouse dihydrofolate reductase. *Nature*, **275**, 617–24.
28. Garapin, A. C., Colbere-Garapin, F., Cohen-Solal, M., Horodniceanu, F. and Kourilsky. P. (1981) Expression of herpes simplex virus type 1 thymidine kinase gene in *Escherichia coli*. *Proc. Natn Acad. Sci USA*, **78**, 815–9.
29. Goeddel, D. V., Heynecker, H. L., Hozumi, T. *et al.* (1979) Direct expression in *Escherichia coli* of a DNA sequence coding for human growth hormone. *Nature*, **281**, 544–8.
30. Hintz, R. L., Wilson, D. M., Finno, J. *et al.* (1982) Biosynthetic methionyl human growth hormone is biologically active in adult man. *Lancet*, **1**, 1276–9.
31. Goeddel, D. V., Yelverton, E., Ullrich, A. *et al.*, (1980) Human leukocyte interferon produced by *E. coli* is biologically active. *Nature*, **287**, 411–6.
32. Goeddel, D. V., Shepard, H. M., Yelverton, E., Leung, D. and Crea, R. (1980) Synthesis of human fibroblast interferon by *Escherichia coli*. *Nucl. Acids Res.*, **8**, 4057–74.
33. Streuli, M., Hall, A., Boll, W. *et al.* (1981) Target cell specificity of two species of a human interferon alpha produced in *Escherichia coli* and of hybrid molecules derived from them. *Proc. Natn Acad. Sci. USA*, **78**, 2848–52.

6 The physical characterization of cloned DNA segments and their counterparts within chromosomes

Genetic engineering techniques provide a means to purify single genes and their associated sequences from complex genomes. The haploid genome of the mouse, for example, contains approximately 3×10^6 kb of DNA. The purification and characterization of a gene such as that encoding β-globin was impossible before the advent of such techniques. The techniques for gene cloning not only enable a gene to be purified but also permit the amplification of the gene during its propagation in *E. coli*. The ability to prepare large amounts of recombinant phage or plasmid DNA facilitates the physical characterization of cloned genes. This chapter will examine the techniques used to study the physical organization of genes. We will see how the information gained goes some way to solving such paradoxes as why most eukaryotes have at least an order of magnitude more DNA than is apparently required to code for their structural genes, and why gene transcripts in the nuclei of most higher organisms are very much larger and more heterogeneous than the functional mRNAs in the cell cytoplasm. Our picture of gene control in eukaryotes is, however, still very incomplete in spite of the information explosion generated by the *in vitro* recombinant DNA technology. This situation should to a large extent be remedied by a combination of *in vitro* mutagenesis (see Chapter 2) with techniques for introducing cloned genes back into eukaryotic cells (see Chapters 7 and 8) in order to study the functional organization of genes. The initial impact of recombinant DNA technology was, however, felt in uncovering the basic features of the physical organization of eukaryotic genes. These basic features will be illustrated in this chapter in the course of describing physical mapping techniques.

6.1 Physical characterization of cloned DNA

6.1.1 Restriction mapping

The distribution of restriction endonuclease cleavage sites on a plasmid or phage DNA molecule can be deduced from the lengths of DNA fragments produced by partial digestion with a single enzyme or by complete digestion with various mixtures of enzymes. The sizes of fragments produced may be determined from their electrophoretic mobilities on gels. Agarose gels are commonly used for double-stranded DNA fragments from about 20 or 30 kb to 100 base pairs. The concentration of agarose in the gel is usually in the range of 0.5% to 2% and is varied according to the size of fragments under investigation. Polyacrylamide gels of 3% to 20% are used for fragments of 1 kb down to six base pairs. The DNA fragments can be conveniently stained with ethidium bromide and visualized by ultra violet illumination. The length of the DNA fragment is inversely proportional to its mobility at low voltage gradients [1]. The size of DNA fragments can therefore be determined from measurements of their mobility relative to DNA fragments of known length.

The commonly used principles for mapping are illustrated for the plasmid pBR322 in Fig. 6.1. The plasmid has unique sites for *Eco*RI, *Pst*I and *Sal*I, and so each of these enzymes alone generates a linear 4.36 kb molecule. A double digest with *Eco*RI and *Sal*I generates fragments of 3.98 kb and 0.38 kb. This positions the *Sal*I site 380 base pairs on one or other side of the *Eco*RI site. This distinction can be made with reference to a third enzyme. *Pst*I, for example, cleaves 750 nucleotides away from the *Eco*RI site. A double digest with *Pst*I and *Sal*I generates fragments of 3.23 kb and 1.13 kb indicating that these sites flank the *Eco*RI site which must be contained within the 1.13 kb fragment. This leads to the circular map shown in Fig. 6.1.

If sites for one enzyme are clustered together without any internal reference cleavage site for a second enzyme then it is usually possible to order the clustered sites from the size of fragments generated by partial restriction cleavage. This problem often arises with enzymes which recognize tetranucleotide sequences and which therefore cleave DNA at frequent intervals. A refinement of this partial digestion technique has been applied to the analysis of repetitive sequence elements in the non-transcribed spacers of tandemly repeating rDNA units. The technique is illustrated in the work of Boseley *et al.* [2] on a cloned *Eco*RI fragment containing the non-transcribed spacer (NTS) and adjacent 28S and 18S sequences of

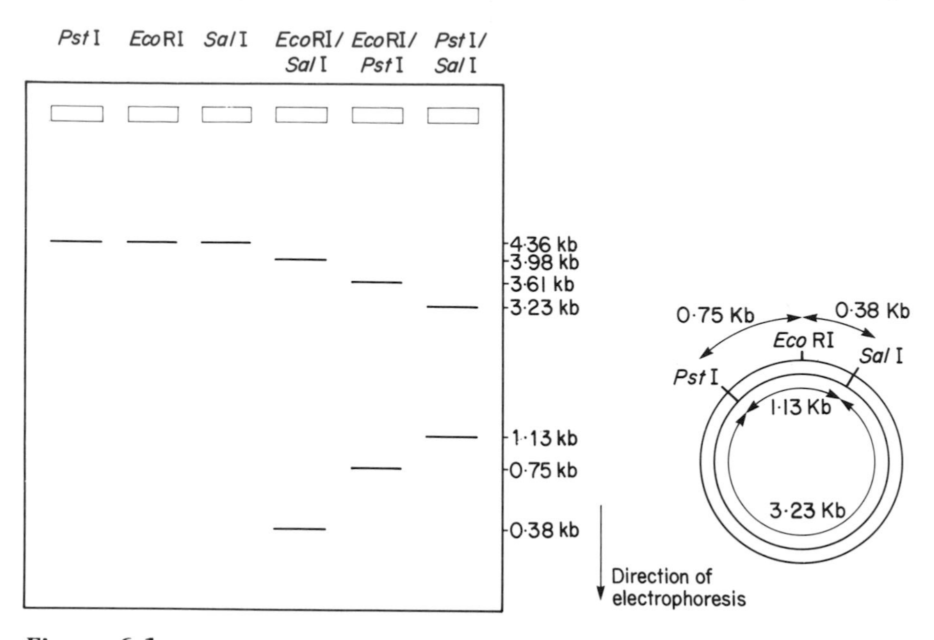

Figure 6.1 Mapping restriction endonuclease cleavage sites on plasmid DNA.

Xenopus laevis rDNA. The first objective is to obtain linear DNA fragments labelled only at one end. The *Eco*RI fragment of rDNA has two internal *Bam*HI sites. The DNA is labelled at its 5′ phosphoryl termini after *Bam*HI cleavage by dephosphorylation using bacterial alkaline phosphatase and rephosphorylation with radiolabelled phosphate from γ^{32}P-ATP using T4 polynucleotide kinase. The terminally labelled *Eco*RI–*Bam*HI fragments contain asymmetric cleavage sites for the enzymes *Hin*dIII, *Hha*I and *Hin*fI, and so cleavage with either of these enzymes generates easily separable left and right 'halves', each of which will then have a terminal label at only one of its 5′ ends. The labelled end is indicated with an asterisk in the upper panel of Fig. 6.2. Partial digestion of these 'half molecules' with an endonuclease for which there are multiple sites will generate an overlapping series of molecules which can be fractionated by gel electrophoresis. A subset of these fragments, which have a common labelled terminus, can be visualized by autoradiography (Fig. 6.2). The order of these labelled fragments and their molecular lengths corresponds directly to the order and location of the restriction sites along the molecule.

The detailed cleavage map of the non-transcribed spacer reveals three repetitious regions: The first region has a repeating unit of 100 base pairs and is separated from region 2 by a non-repetitive element in which is found a *Bam*HI site. This non-repetitive element has been termed a '*Bam* island'. A similar non-repetitive element

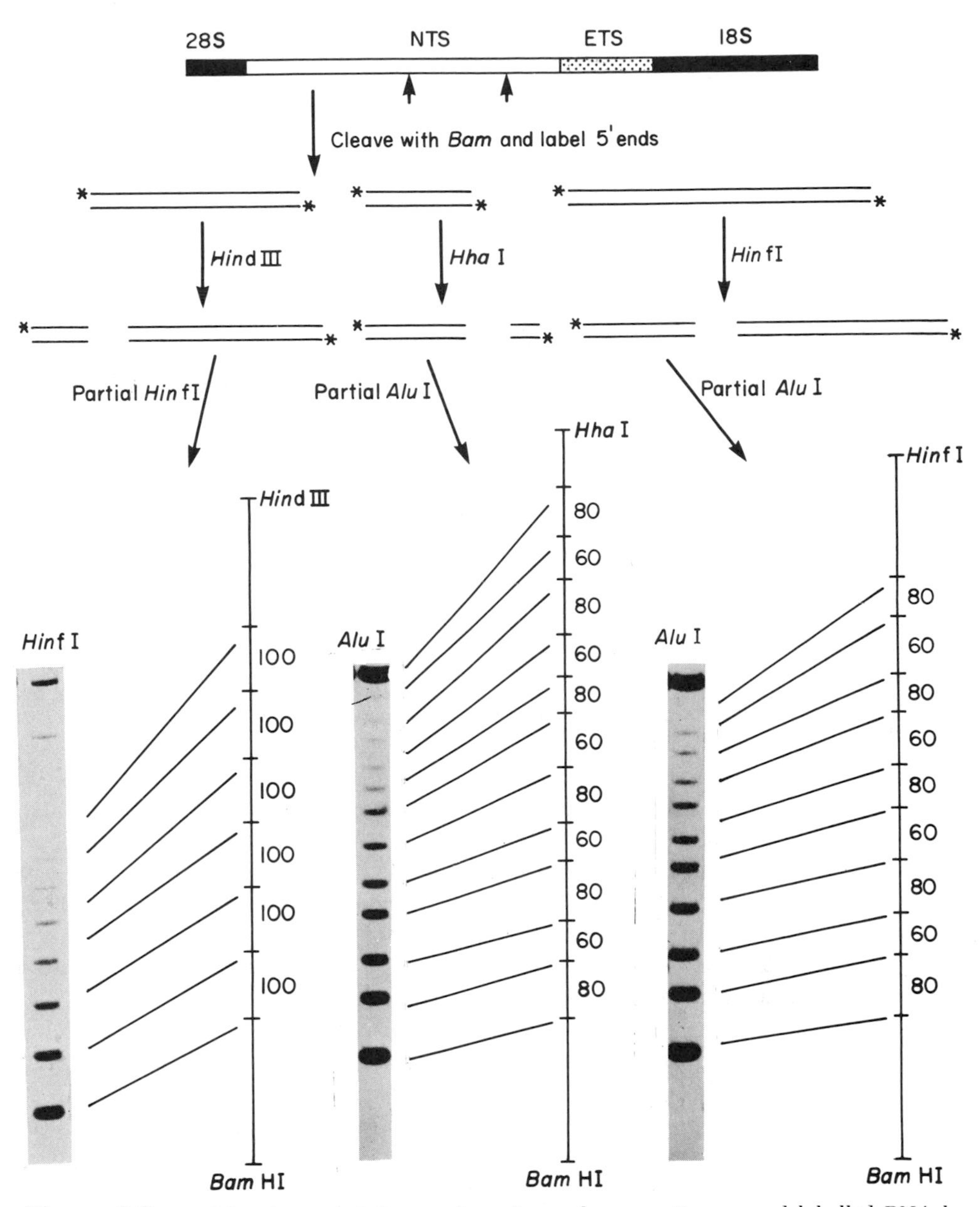

Figure 6.2 Mapping restriction endonuclease cleavage sites on end labelled DNA by partial digestion (data taken from [2]).

separates regions 2 and 3. These two regions are virtually identical, having alternating 81/60 base pair arrangements, with region 3 differing from region 2 in having *Sma*I sites in the 81 base pair unit. DNA sequencing of the units in regions 2 and 3 shows that the 60 and 81 base pair arrangements are identical, excepting a run of 21 nucleotides. There is also some homology between the nucleotide sequence of the '*Bam* island' and repeating units 1, 2 and 3 and between the '*Bam* island' sequence and the putative promoter region to the 5′ side of the junction between the non-transcribed spacer and the external transcribed spacer (ETS). A hypothetical scheme can be constructed from these data to account for the evolution of the non-transcribed spacer by a series of reduplication events upon an ancestral sequence postulated to be at the NTS/ETS boundary. This leads to the interesting idea that, by duplicating what is potentially a promoter region, a DNA sequence has been generated which has the capability of binding many RNA polymerase molecules.

6.1.2 Mapping cloned DNAs by electron microscopy

A homogeneous population of DNA molecules can be characterized directly by observation in the electron microscope. The molecules can be photographed and measured and so identifiable physical features can be mapped. These features include regions that are AT rich and melt before the rest of the molecule; regions of homology between different molecules identified by their capability to form duplex structures; and regions with homology to RNA molecules. In order to visualize DNA in the electron microscope (EM), a basic protein, usually cytochrome c, is added to the DNA solution and the mixture allowed to run down a glass ramp and form a molecular film over a liquid surface (the hypophase). The nucleic acid–cytochrome c complex is then picked up onto the surface of an electron microscope grid coated with a film of parlodion (a form of nitrocellulose). This is done simply by touching the coated grid onto the molecular film formed by the DNA–cytochrome solution (Fig. 6.3). The grids are stained in an ethanolic solution of uranyl acetate and subsequently subjected to low angle rotary shadowing with a platinum–palladium mixture. In general, two types of spreading technique are used. The aqueous technique (in which the hypophase and the solution containing the DNA are of ammonium acetate) allows the visualization of double-stranded nucleic acids only; the single-stranded nucleic acids collapse into a bush. In another technique both single-stranded and double-stranded nucleic acids can be visualized when the DNA is spread over buffered formamide solutions.

(*a*) *Denaturation mapping*

This technique permits the mapping of AT-rich regions within a DNA molecule. Such regions are the first to melt when double-stranded DNA is exposed to denaturing conditions. A partially denatured DNA molecule can be visualized by spreading at very high formamide concentrations. Alternatively the molecule can be partially denatured with alkali and the single-stranded regions fixed with formaldehyde prior to spreading in a moderate concentration of

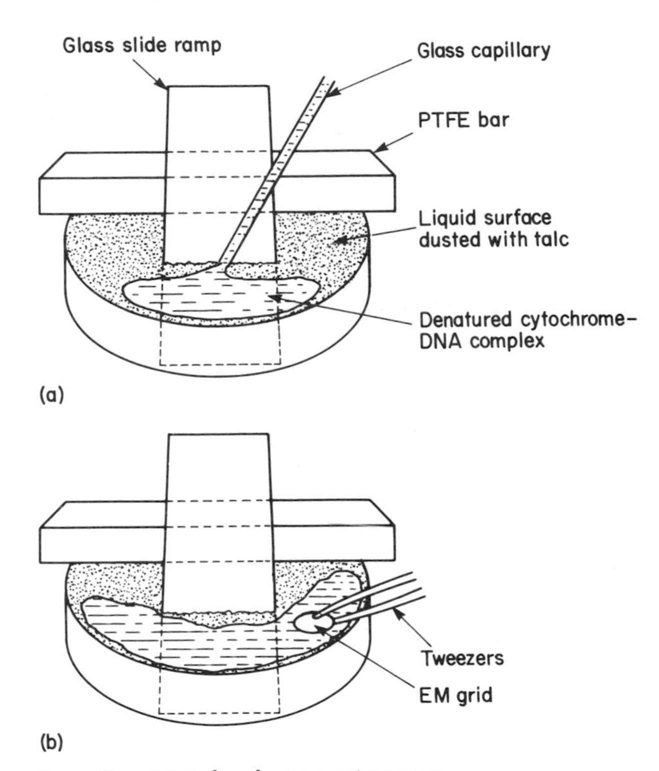

Figure 6.3 Spreading DNA for electron microscopy.

formamide such as is normally used to prevent the collapse of single-stranded regions. The example shown in Fig. 6.4 is a cloned segment of *Drosophila melanogaster* DNA which contains 32 tandem repeats of the gene for 5S rRNA cloned in the plasmid ColEl [3]. The plasmid vector sequences to the left of the micrograph contain a long stretch of GC-rich DNA, on each side of which are two large denaturation bubbles. To the right of the molecule the tandemly repeating 5S DNA forms a regular array of double-stranded sequences corresponding to the GC-rich gene, and small denaturation bubbles corresponding to the AT-rich spacer.

(b) Heteroduplex mapping

This is probably the most widely applied EM mapping technique permitting the localization of regions of sequence homology between DNA molecules. The two types of DNA molecule are mixed and denatured in alkali. The solution is neutralized and formamide added to allow reassociation at room temperature. Following reannealing for such a time that permits about 50% of homologous sequences to form duplex structures, the mixture is diluted and spread.

In the example shown in Fig. 6.5, heteroduplex mapping is used to

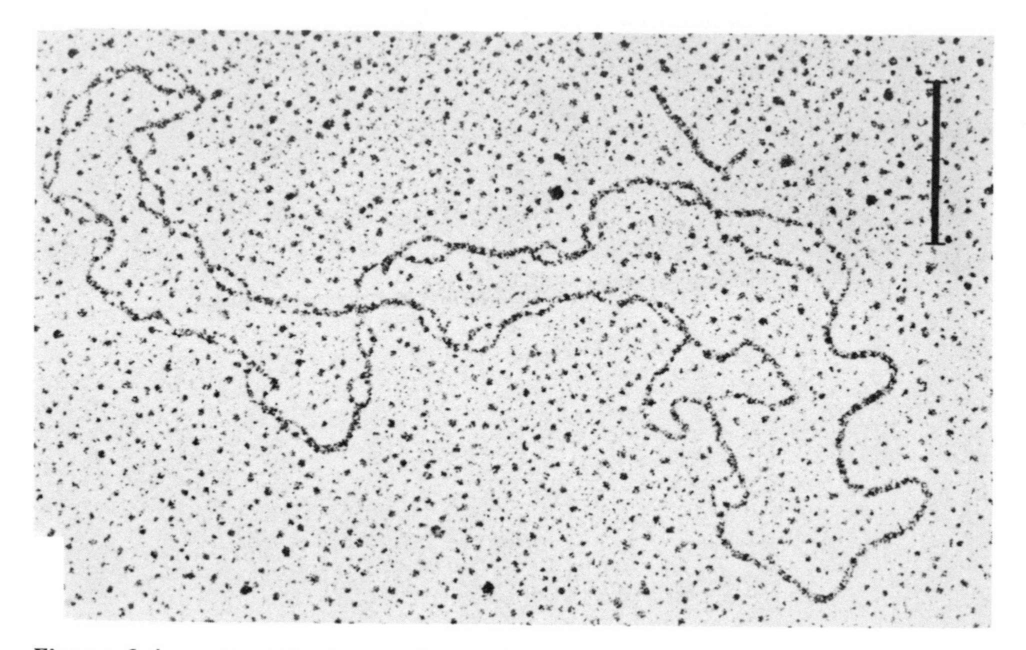

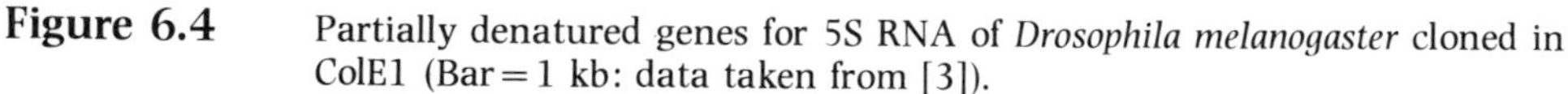

Figure 6.4 Partially denatured genes for 5S RNA of *Drosophila melanogaster* cloned in ColE1 (Bar = 1 kb: data taken from [3]).

localize homologous regions between two members of a gene family which occur at different chromosomal sites. These are the genes for the α and β subunits of the larval serum protein 1 (LSP1) of *Drosophila melanogaster* [4]. This is the most abundant protein made in third instar larvae and is thought to be a storage protein utilized in metamorphosis. The α and β genes are cloned in bacteriophage λ and the phage arms are annealed to give double-stranded DNA. The α and β LSP1 genes are located on the X chromosome and chromosome 2 respectively. The *Drosophila* DNA flanking the genes is therefore not homologous and remains single-stranded, whereas the regions of homology between the two genes form duplex DNA.

(c) *Mapping regions homologous to RNA*

A number of techniques have been developed to look at RNA annealed to DNA. The most straightforward is simply to anneal denatured DNA with RNA and then examine the structure following formamide spreading. The interpretation of such DNA–RNA duplexes can be very difficult if one has annealed a linear DNA molecule, containing a gene plus flanking sequences, to an RNA molecule perfectly colinear with the gene, since it is visually difficult to position the junction between single-stranded DNA and DNA–RNA duplex.

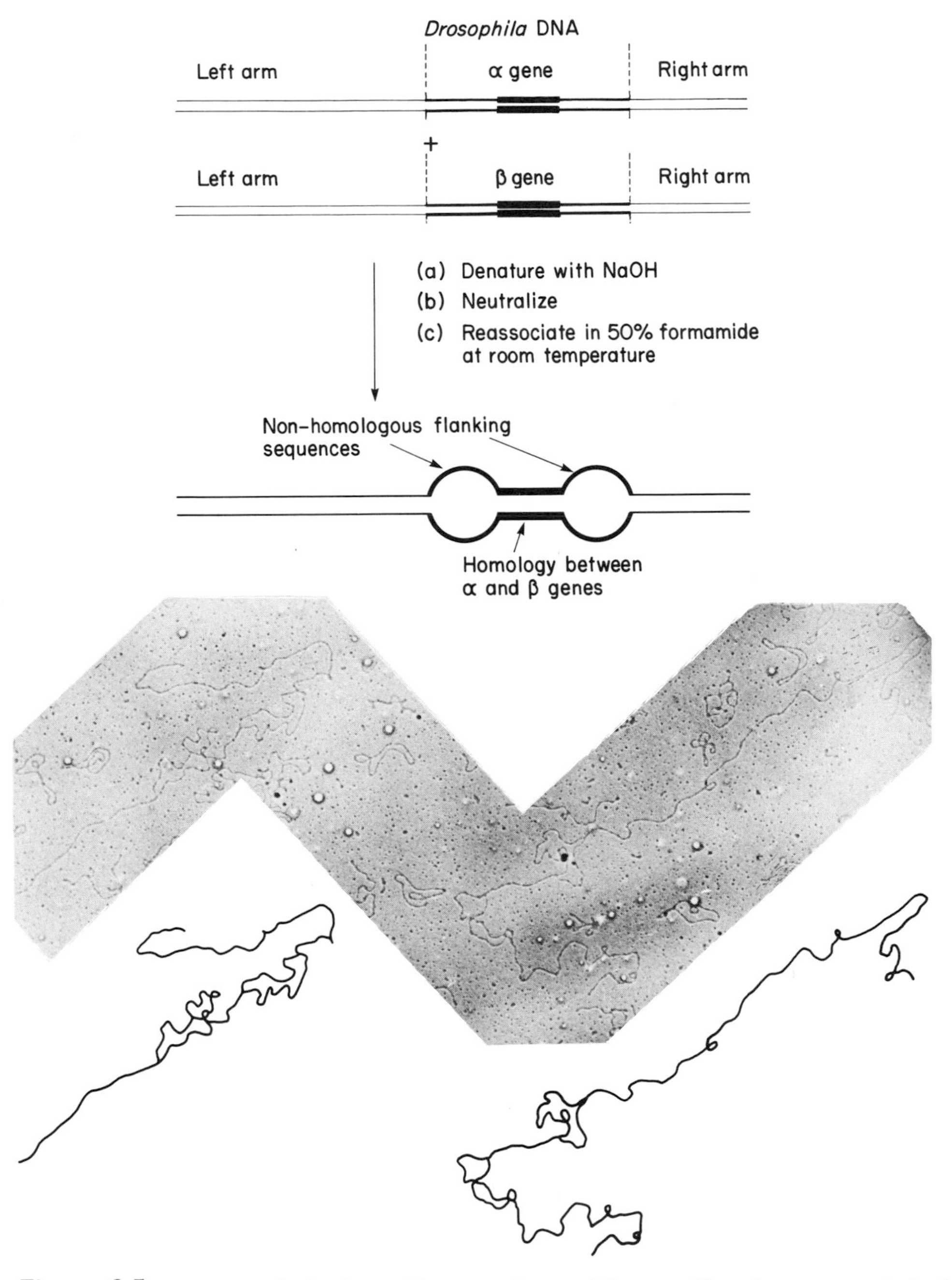

Figure 6.5 Heteroduplex formed between the α and β genes of larval serum protein 1 of *Drosophila melanogaster* cloned in bacteriophage λ (data taken from [4]).

This is less of a problem with a eukaryotic gene containing intervening sequences which will form loops of single-stranded DNA at the splice positions on the mature RNA molecule. Indeed the technique has been extensively used by Chambon and his coworkers to map the positions of intervening sequences on several genes from the chicken [5].

There are two other approaches which are extensively used to map the endpoints of genes which do not have intervening sequences. Fig. 6.6 is a micrograph of messenger RNA for three of the five sea

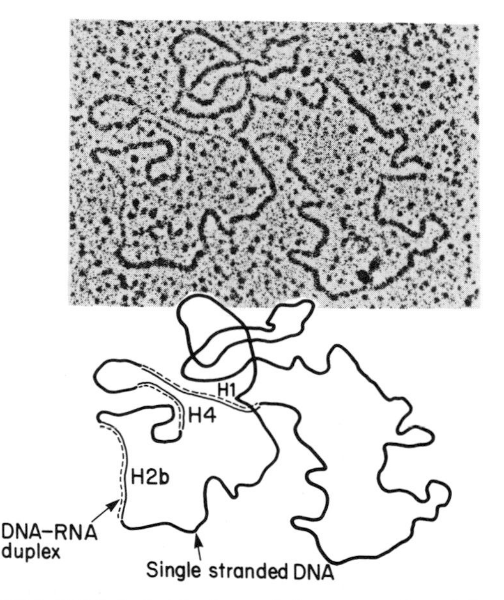

Figure 6.6 DNA–RNA hybrids formed between cloned histone genes from the sea urchin, *Strongylocentrotus purpuratus* and their mRNAs examined by the gene 32 protein–ethidium bromide spreading technique (data taken from [6]).

urchin histones annealed to a segment of the histone genes and spread in the presence of T4 gene 32 protein and ethidium bromide [6]. The gene 32 protein binds specifically to single-stranded DNA and as a consequence the single-stranded regions appear fatter than double-stranded regions. The addition of the intercalating dye, ethidium bromide, to spreading mixtures gives double-stranded DNA a smoother contour. The five histone genes separated by spacer sequences are arranged as tandemly repeating units. Similar tandem arrangements are found in other organisms, although the order and relative orientation of the five genes within the repeating unit are not necessarily identical to that found in the sea urchin.

The second approach is that of R-loop mapping [7], which also makes use of the increased stability of DNA–RNA duplexes in high formamide concentrations. Unlike the S1 mapping technique discussed later in this chapter, the annealing is carried out with RNA and double-stranded DNA at a temperature which is ideally 1° C below that required for complete DNA–DNA strand separation. Under these conditions, the RNA anneals with its complementary sequence within the DNA molecule and displaces a single strand of DNA to give the R-loop (Fig. 6.7).

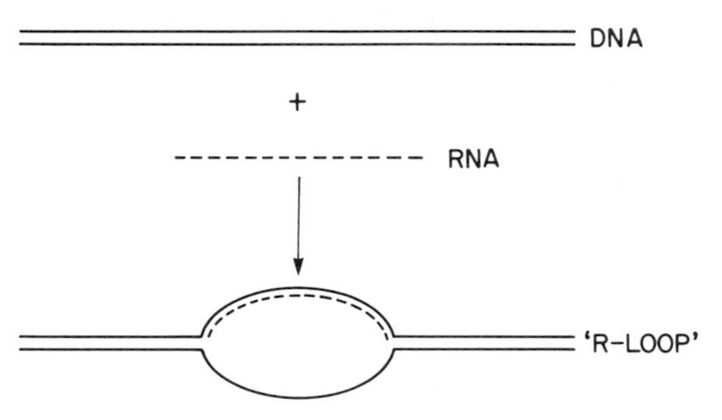

Figure 6.7 The formation of R-loops.

Fig. 6.8 shows R-loop mapping applied to the mouse genes for the immunoglobulin $\lambda 1$ light chains [8]. The characterization of cloned segments of these genes, and the application of gel transfer hybridization studies on total genomic DNA with the cloned segments as probes, has shown that the DNA coding for the variable (V) and constant (C) regions of the polypeptide is widely separated in embryonic cells but is brought together by specific recombinational events in differentiated cells which produce immunoglobulin. In embryonic DNA the sequence that encodes the major part of the V region (amino-acid residues 1–97) is on one DNA segment and the remainder, termed the J region (coding for residues 98–109), is on a separate DNA segment. Clones containing part of the V region give simple R-loop structures consisting of a single loop in the region of homology between DNA and mRNA, with the 3′ end of the RNA molecule hanging free. The J segment is closely linked to the C region of the gene, but the two are separated by a 1.2 kb intervening sequence. Intervening sequences were first found in a number of eukaryotic genes as a result of EM studies such as this carried out on cloned DNA molecules and by gel transfer hybridization experiments (see Section 6.2.2(a)). The intervening sequence is initially transcribed into RNA, but subsequently removed from the transcript by a splicing process. The R-loop structure formed by the cloned segment containing the J and C regions shows two loops which correspond to DNA–RNA hybrid and displaced single-stranded DNA, between which

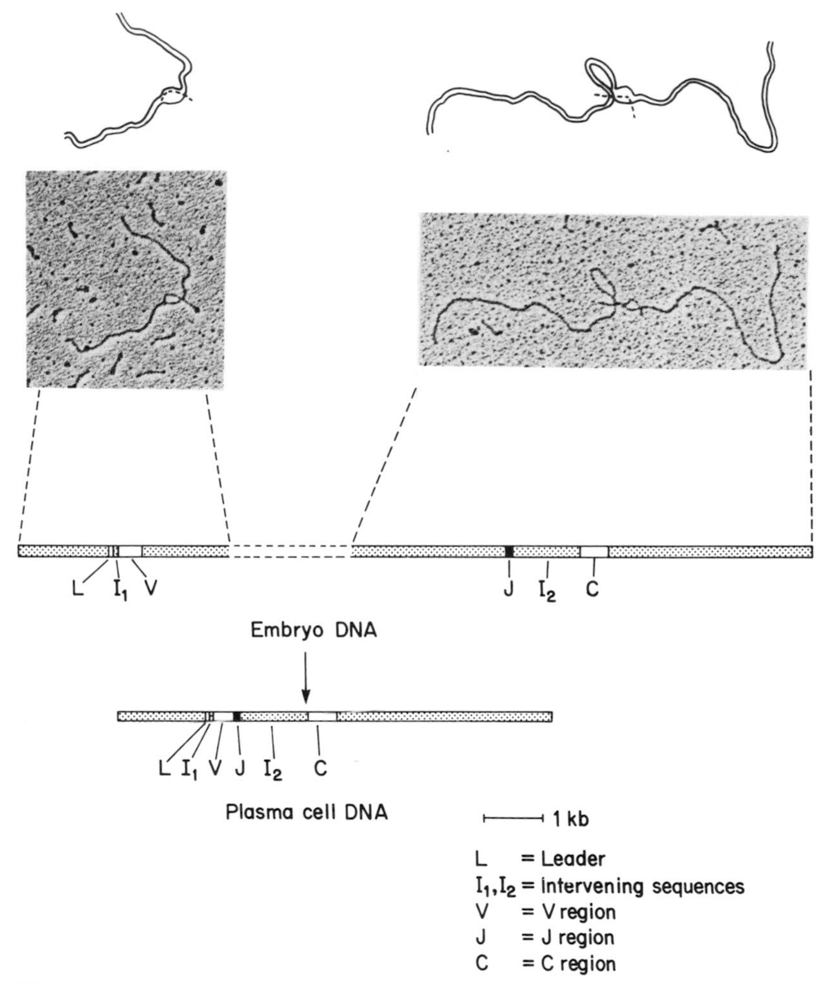

Figure 6.8 R-loops formed by the cloned λ-immunoglobulin genes of mouse (data taken from [8]).

is a loop of double-stranded DNA corresponding to the intervening sequence. At the end of the structure which has the smaller R-loop is a tail of the unannealed 5′ end of the mRNA and at the 3′ end of the gene a whisker of the poly-A sequences. In DNA from the differentiated cells the 3′ end of the V region is linked to the 5′ end of the J region and a cloned segment of this DNA forms two R-loops representative of the full length of the mRNA but still retaining the intervening sequences between the J and C regions. These sequence relationships (shown diagrammatically in Fig. 6.8) have been confirmed in heteroduplex experiments between the cloned DNAs.

The above experiments address V–C joining for the λ light chain genes of mouse. Additional diversity is generated for the κ light chain genes by the presence of multiple J segments, any one of which can be brought into juxtaposition with the C segment by V–C joining. Similar somatic recombination events take place during the differentiation steps resulting in functional heavy chain genes. This brings a V_H region into close proximity to the DNA encoding the larger three-domain C_H region. In this case there are multiple copies of yet another element, the D segment, such that finally the gene is comprised of four components, V_H, D, J_H, and C_H. A final level of recombinational activity occurs within the differentiated cell, that enables the cell to make heavy chains with C_H regions of different types whilst keeping the same V_H region. During this process of 'heavy chain switching' the V_H coding region established by the recombinational events of differentiation is transferred from one C_H coding region to another. Lucid accounts of the overall organization of the immunoglobulin gene families and of the contribution made by both somatic recombination and somatic mutation to the generation of antibody diversity can be found in references 9 and 10.

6.1.3 Electrophoretic techniques for mapping transcripts

(*a*) *S1 mapping*

EM techniques can position a physical feature on a DNA molecule to within 50–100 base pairs. An alternative set of electrophoretic techniques are available for mapping regions homologous to RNA which can localize the end of such a region to the precise nucleotide pair. One of these approaches was developed by Berk and Sharp [11] for mapping viral transcripts. The technique rests upon the observation that in high concentrations of formamide DNA–RNA duplexes are more stable than the equivalent DNA–DNA duplex [12]. It is possible to thermally denature DNA duplex under such conditions and then lower the temperature to a point at which DNA and RNA will anneal, but at which DNA will not reassociate with itself. DNA which remains single-stranded can then be digested with S1 nuclease and the size of the resulting DNA–RNA duplex determined by electrophoresis on an agarose gel. If the DNA has first been cleaved with a restriction endonuclease that cleaves within the coding region, the length of DNA protected from S1 digestion by RNA will correspond to the distance between the end of the RNA coding region and the restriction endonuclease cleavage site. The principles of this approach are shown in Fig. 6.9 for a hypothetical eukaryotic gene which contains one restriction endonuclease cleavage site and an intervening sequence. The mature mRNA is shown annealed to complementary sequences in one of the DNA strands. S1 nuclease

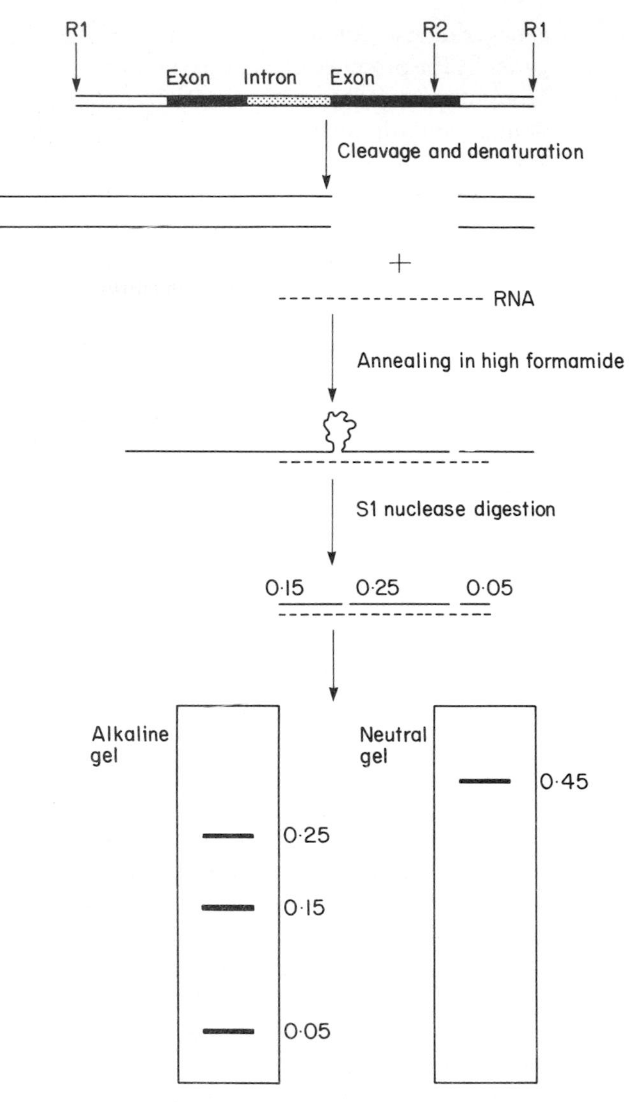

Figure 6.9 The principles of mapping transcripts using S1 nuclease.

acts upon this hybrid structure to cleave single-stranded DNA preferentially. The intervening sequence DNA is therefore digested by the S1 nuclease and the annealed mRNA holds the two flanking exons together. The position of the intervening sequence can then be deduced from a comparison of the electrophoretic mobility of this DNA–RNA hybrid on a neutral gel or on an alkali gel in which the RNA is hydrolysed and the two exon fragments migrate independently.

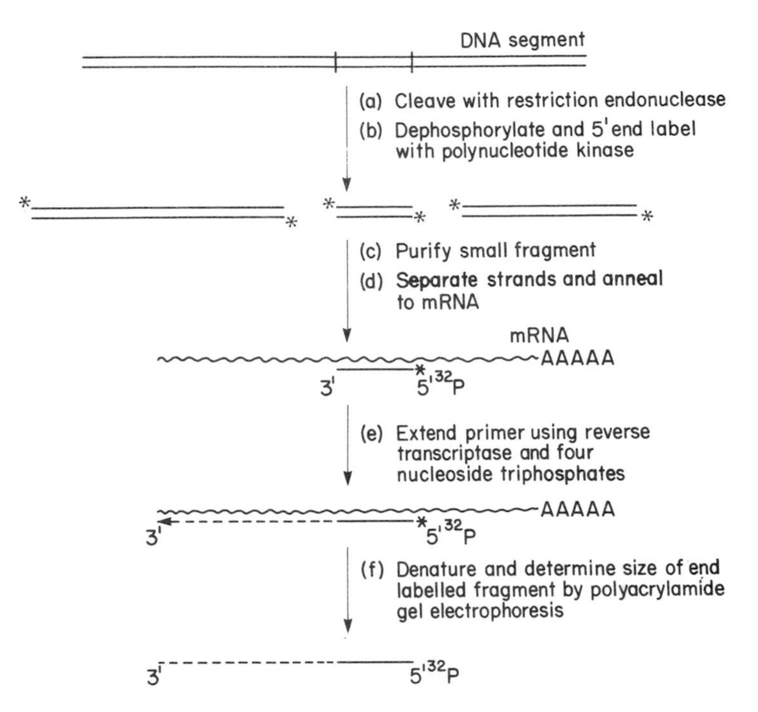

Figure 6.10 Mapping transcripts by primer extension.

(*b*) *Primer extension*

An alternative approach which has been used to localize the 5′ end of an RNA molecule relative to restriction endonuclease cleavage sites in DNA, uses a restriction fragment close to the 5′ end of the message to prime the synthesis of cDNA. The principles are illustrated in Fig. 6.10. The restriction fragment is allowed to anneal with the RNA which then serves as a substrate for reverse transcriptase. The restriction fragment is extended by the incorporation of complementary deoxynucleotides until the 5′ end of the mRNA is reached. The reaction is usually carried out using end-labelled primer DNA. The length of the extended primer can then be determined by electrophoresis on a 'sequencing gel' alongside known markers.

6.2 Detecting specific nucleic acid sequences within heterogeneous populations of molecules – gel transfer hybridization

6.2.1 DNA blotting

A restriction endonuclease which recognizes a hexanucleotide sequence will cleave a random nucleotide sequence once every 4096

(4^6) base pairs. Wild type phage λ DNA (49 kb), for example, has five cleavage sites for *Eco*RI and six sites for *Hin*dIII, and so the cleavage sites in such a DNA molecule can be readily mapped by direct electrophoretic analysis of the cleavage products following ethidium bromide staining. Mammalian genomes, however, will be cleaved into approximately 10^6 specific fragments by either of these enzymes, resulting in a smear of fragments that cannot be resolved by electrophoresis. A method has been developed by Southern [13] which permits the detection of specific sequences within this smear and consequently the restriction mapping of those sequences. The principles of the method are shown in Fig. 6.11. Following electrophoretic fractionation, the fragments of DNA are denatured within

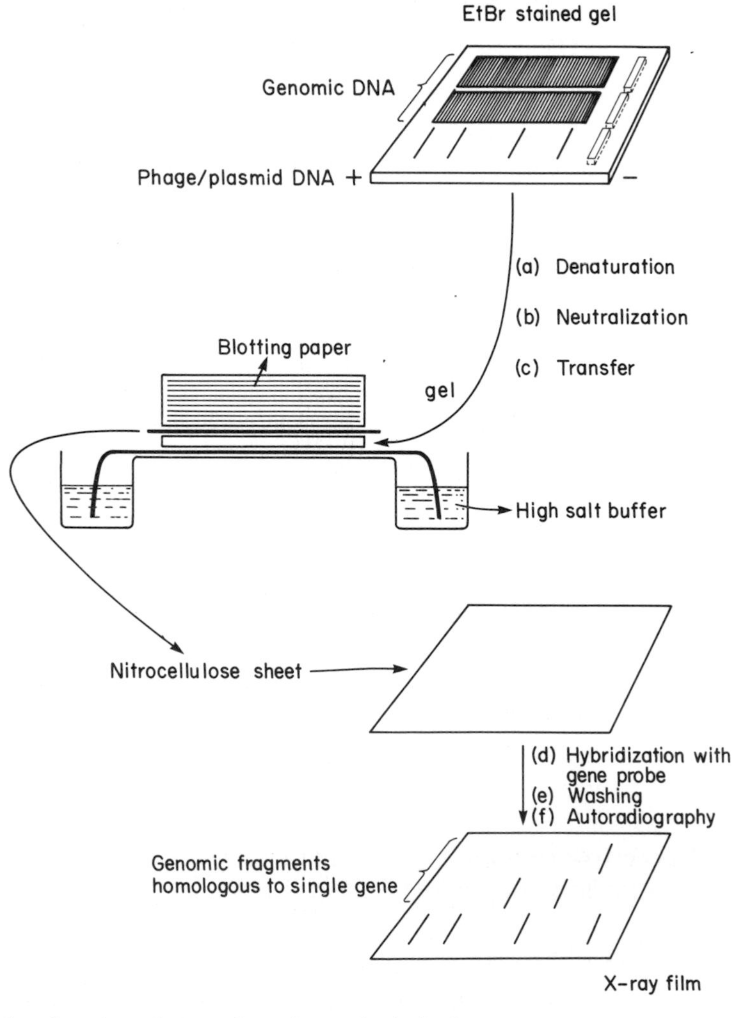

Figure 6.11 Southern's technique for gel transfer hybridization.

the gel and then 'blotted' onto a sheet of nitrocellulose to which they bind retaining the same relative positions that they occupied within the gel. The nitrocellulose filter is then incubated together with a ^{32}P-labelled sequence probe, which will hybridize to its complement within the smear of restricted DNA. Unhybridized probe can then be washed away and the position of hybridization on the nitrocellulose detected by autoradiography. Autoradiographic techniques [14] can detect as little as one disintegration of ^{32}P per minute (dpm). DNA can be labelled *in vitro* to greater than 10^8 dpm μg^{-1} by nick translation [15] and so it is clearly possible to use this technique to detect unique restriction fragments which represent approximately one part in 10^6 of a digest of mammalian DNA. This technique has found widespread application for the analysis of genomic sequences wherever cloned segments are available for use as radiolabelled probes.

The power of the technique is illustrated by the early studies of Jeffreys and Flavell [16] on the structure of the rabbit β-globin gene in chromosomal DNA. These workers used an almost complete cDNA copy of the rabbit β-globin mRNA [17] as a hybridization probe to detect the corresponding chromosomal sequences. The smear of ethidium bromide stained chromosomal fragments is shown for duplicate digests of rabbit DNA with *Eco*RI and *Hae*III in Fig. 6.12. The autoradiograph given by this same DNA after blotting onto nitrocellulose and hybridization with the cDNA probe is shown alongside the stained gel. A restriction endonuclease cleavage map can be constructed from such autoradiograms following the approaches described earlier in this chapter. A detailed restriction analysis of the chromosomal β-globin sequences showed them not to be present in a continuous stretch of DNA, but instead to be in at least two pieces separated by a 600 base pair non-β-globin sequence. This conclusion was based on the mapping of intragenic sites for *Bam*HI, *Taq*YI and *Hae*III 600 base pairs too far to the left of an intragenic *Eco*RI site by comparison with the map of the cDNA clone, and also on the fact that the enzyme *Hap*II cuts the chromosomal gene in half whereas there is no *Hap*II site within the cDNA (Fig. 6.12). Such discontinuities caused by intervening sequences have subsequently been found to be common organizational features of eukaryotic genes.

6.2.2 RNA blotting

A second type of gel transfer hybridization has been developed by Alwine *et al.* [18] in order to analyse the distribution of specific RNA sequences within a population of electrophoretically fractionated RNA molecules. The experimental approach is identical in Southern's technique, but a different solid-phase support is used to

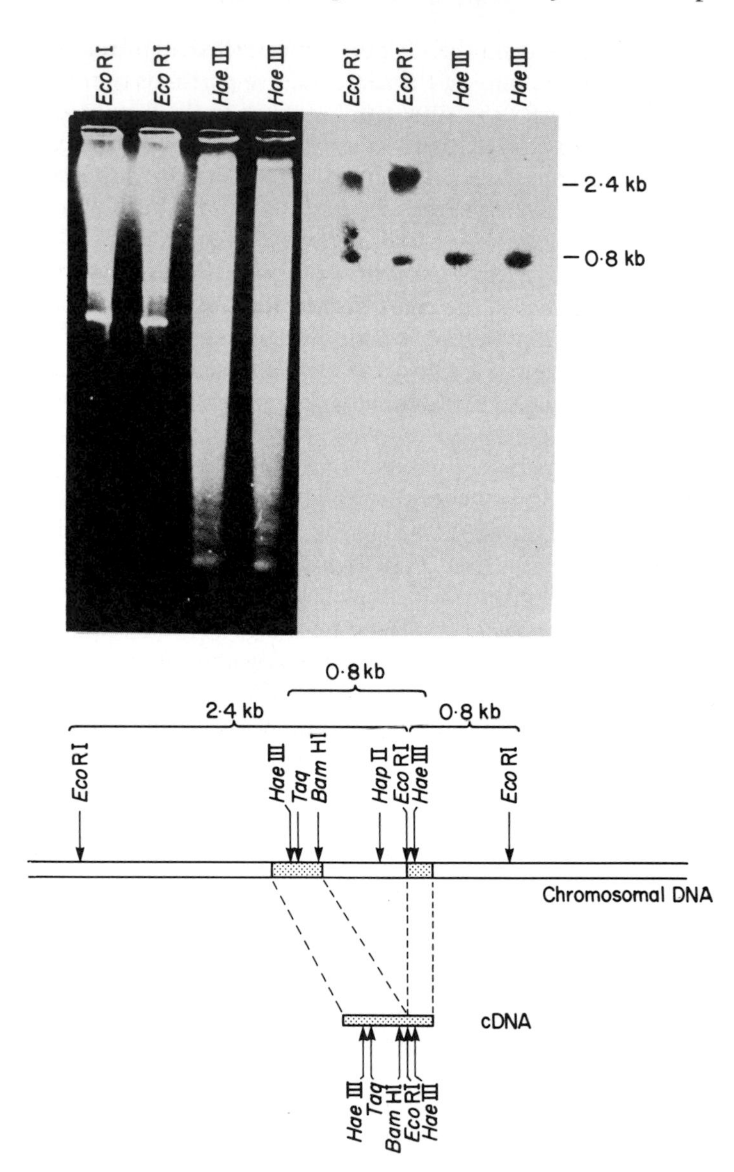

Figure 6.12 Mapping the β-globin gene in rabbit DNA by 'Southern blotting' (from [16]).

bind and immobilize RNA in the position to which it has migrated on a gel. Ed Southern seems to be resigned to the term 'Northern' blotting which has been coined to describe this technique. The support originally used is paper derivatized with diazobenzyloxymethyl (DBM) groups to which single-stranded nucleic acid will bind covalently. It has subsequently been shown that nitrocellulose sheets will bind RNA that is fully denatured. This is also important to achieve a linear inverse relationship between the mobility of an RNA molecule and its molecular length. The gels are therefore run in the presence of methylmercury, glyoxyl or formaldehyde as denaturing agents. Following its transfer to DBM paper or nitrocellulose, the immobilized RNA can be hybridized with a radiolabelled DNA probe. This approach has found widespread application in the analysis of the transcripts of cloned eukaryotic genes. Fig. 6.13 shows the result of such an experiment in which nuclear and cytoplasmic RNAs from the cells of the chick oviduct have been blotted onto DBM paper and probed with the radiolabelled chicken ovalbumin gene. Ovalbumin is the major egg white protein. Its gene, which contains seven intervening sequences, is expressed in the chick oviduct under

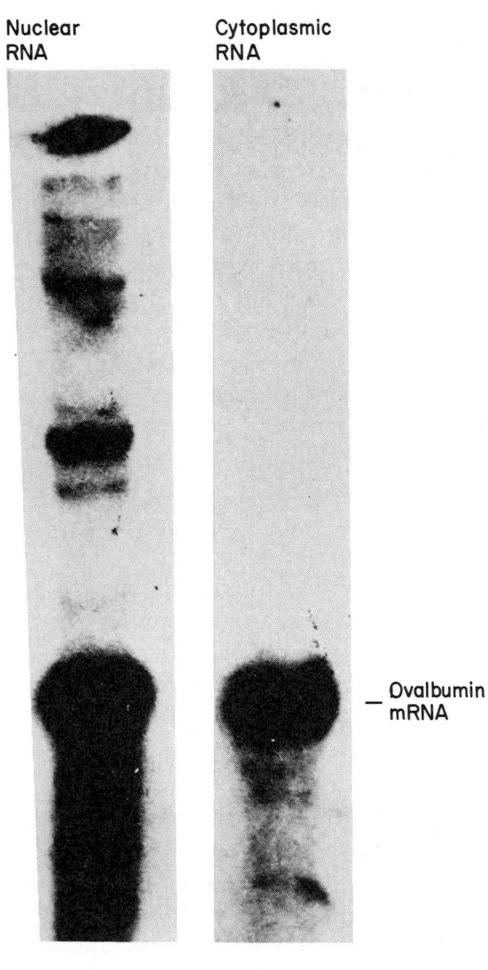

Figure 6.13 Hybridization of electrophoretically fractionated nuclear and cytoplasmic RNA from chick oviduct cells with the cloned ovalbumin gene (from [19]).

stimulation by oestrogen. As seems to be the case for most genes which contain intervening sequences, the entire chromosomal sequence is transcribed and the intervening sequences are subsequently removed from the primary transcript. The coding regions of RNA are then spliced together to give the mature message. Fig. 6.13 shows the mature cytoplasmic 2 kb ovalbumin mRNA. The nuclear RNA contains seven additional transcripts of the gene which range up to greater than four times the length of the mRNA [19]. These RNAs represent processing intermediates from which the intervening sequences are successively removed.

6.3 Mapping cloned DNA to chromosomal loci

6.3.1 *In situ* hybridization

The most direct way of localizing a gene to a chromosomal site is to hybridize a cloned DNA sequence to a chromosome preparation *in situ*. The chromosome preparation is immobilized on a glass slide and the DNA in the chromosomes denatured. The preparation is then incubated with a probe of ^{3}H-DNA or ^{3}H-RNA under conditions that enable the probe to anneal with its complementary sequences in the chromosomes. The unhybridized material is washed away and the position of hybridization localized by autoradiography. This problem has been most successfully applied to the polytene chromosomes of *Drosophila melanogaster*. This organism has four pairs of chromosomes which in certain tissues replicate to give ploidies of up to 1056. The chromatids do not separate but become laterally aggregated to form the giant chromosomes. The centromeric sequences do not undergo as many rounds of replication as do the rest of the chromosomal sequences and in the polytene state they aggregate to form the heterochromatic chromocentre. The homologous arms of each chromosome pair also aggregate so that a polytene squash has the appearance of five long chromosomal arms radiating out from the chromocentre. These arms have a characteristic banding pattern and there is a long established correlation between genetic and cytological chromosome maps. The endomitosis of *D. melanogaster* chromatids to form the polytene chromosomes ensures that a gene which may be described as 'unique' within the haploid genome is present at one chromosomal location in sufficient copies to permit its detection by *in situ* hybridization with low specific activity ^{3}H probes. An example of an *in situ* hybridization experiment with *D. melanogaster* salivary gland chromosomes is shown in Fig. 6.14. The probe in this experiment is a cloned segment of DNA, cDm412, which encodes a 7 kb polyadenylated RNA. Dm412 is an example of a 'copia-like' transposable element and resembles the retro-viruses of the higher eukaryotes (see Chapter 8) in many respects. Dm412 is not found at the same sites in all strains of *D. melanogaster*. This is demonstrated in the autoradiogram (Fig. 6.14), which has been carried out on the chromosomes from the F1 progeny formed by crossing the Seto strain with the Oregon-R strain [20]. In the asynapsed region of the second chromosome of the squash shown in Fig. 6.14, none of the hybridizing sites are common to both the parental chromosomes.

The low specific activity of ^{3}H labelled probes makes it difficult to utilize the technique to localize sequences within mitotic

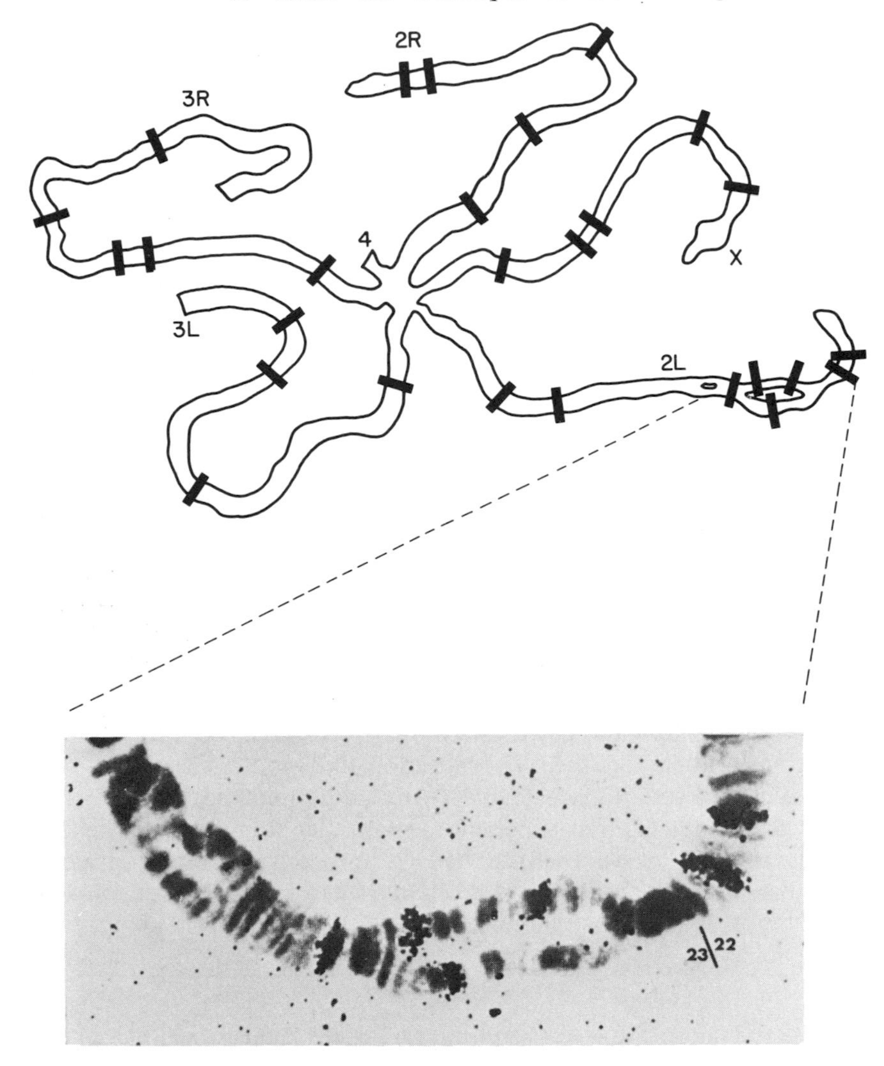

Figure 6.14 The hybridization of Dm412 sequences to multiple sites on the polytene chromosomes of *Drosophila melanogaster in situ* (from [20]).

mammalian chromosomes. It is however still possible to do this. Only a few grains are seen over any one set of metaphase chromosomes and many of these represent background signal. If the total number of grains is taken for many spreads, however, then the result is statistically meaningful. Several genes have been mapped in this way, including the human insulin gene [21] and interferon genes [22]. Fig. 6.15 shows the results of an *in situ* hybridization with a segment of DNA from an X chromosome gene library [25]. Five silver grains are seen over the region Xq22–23.

6.3.2 Somatic cell genetics

Southern blot hybridization techniques have been used in conjunction with somatic cell genetics to localize cloned mammalian genes to chromosomal regions. The principles of somatic cell genetics are based on techniques whereby cultured cells from different species can be fused together using Sendai virus or polyethylene glycol. The hybrid cell lines that are formed have a tendency to lose chromosomes from one of the parents. In hybrids formed between mouse and human cells, for example, the human chromosomes are progressively eliminated. It is possible to take hybrid cell lines in which a small number of human chromosomes are stably retained and analyse their DNA for sequences complementary to a cloned gene. This approach was first used to map the human α-globin gene cluster to chromosome 16 and the β-globin gene cluster to chromosome 11 [23]. A more detailed map position can be obtained by using chromosome translocations. There are more than 300 human reciprocal translocations that have been mapped, in which segments of different chromosomes have been exchanged. The individuals carrying such chromosomes suffer from congenital abnormalities. Cell lines established from these individuals can be fused with mouse cells as described above. The hybrid cell lines that emerge from such a procedure in some cases retain one of the human chromosome translocations. Southern hybridization analysis of such cell lines can localize a gene to one of the two chromosomal segments in the translocation. In experiments to localize the human insulin gene a human cell line was used that had a reciprocal translocation between chromosomes 11 and 14. This cell line was fused to mouse cells, and

---

Figure 6.15 Hybridization of a cloned segment of human X chromosome DNA to the region Xq22–23. The lower panel shows the G-banded metaphase spread. The upper panel shows the same spread after hybridization and autoradiography (unpublished photographs of David Hartley).

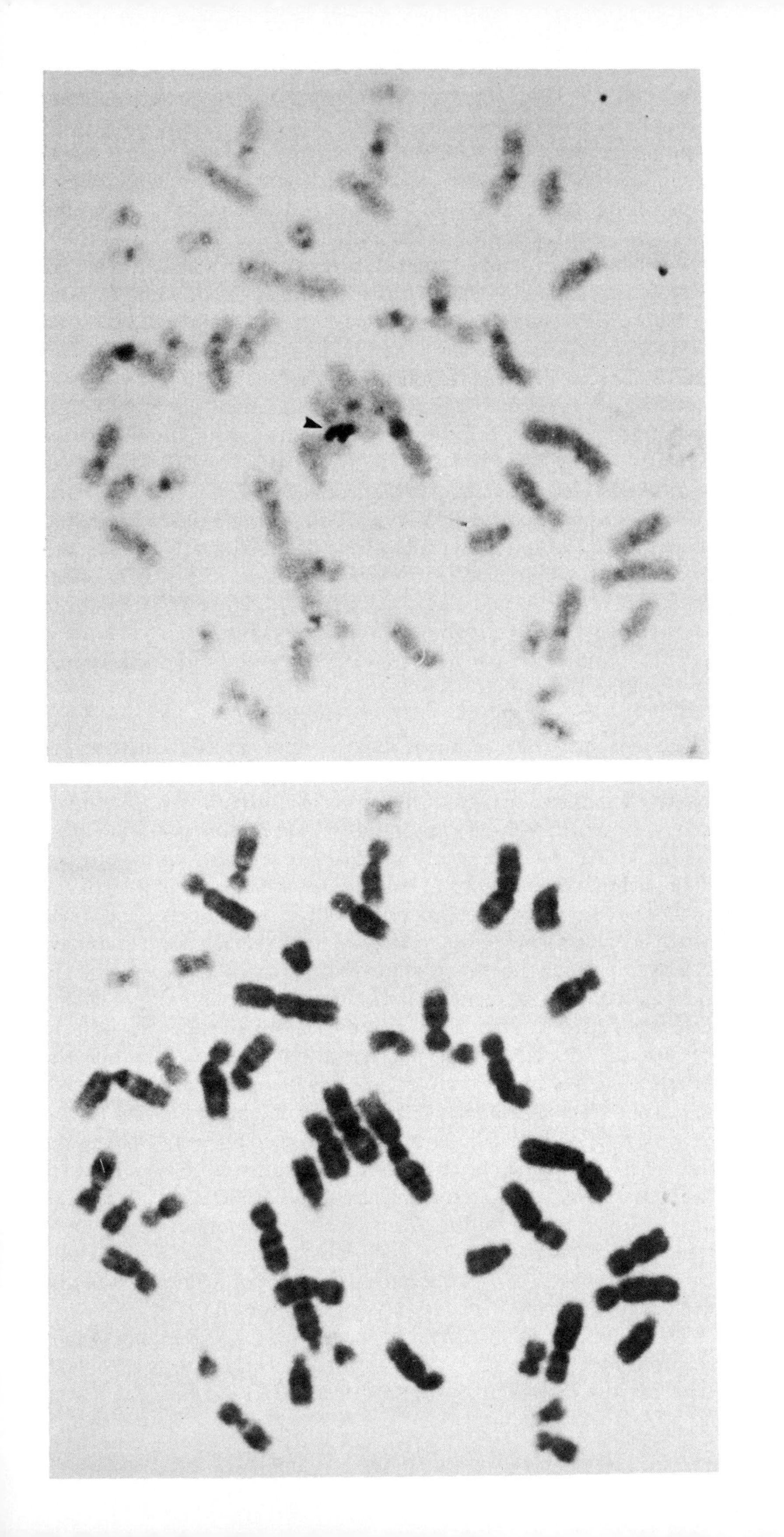

hybrids were isolated which retained either the 11 to 14 translocation or the 14 to 11 translocation. Southern hybridization experiments on the DNA from these two cell lines localized the insulin gene on the short arm of chromosome 11 [24].

Hybrid cell lines that retain a single human chromosome are excellent starting material for the cloning of DNA from a single human chromosome. A library of recombinant DNA molecules can be made from such DNA using the techniques described in Chapters 2 and 3. Such a library can then be screened for human specific sequences. An alternative approach that has found success involves staining preparations of mitotic chromosomes with a fluorescent dye and then using a fluorescence activated cell sorting machine to separate the chromosomes on a basis of their size. The cells which are chosen must have a suitable karyotype to facilitate the fractionation of particular chromosomes. It has been possible, starting with a cell line containing four X chromosomes, to prepare enough X chromosomal DNA to build a phage library [25]. The cloned probe for the *in situ* hybridization experiment shown in Fig. 6.15 is from this library.

6.4 Genetic diseases

There are over 500 heritable diseases that result from recessive mutations. In many cases the primary genetic defect for the disease has not been identified. Cystic fibrosis is one such disease, the genetic defect for which is carried by roughly one person in twenty. One couple in 400 have therefore a 1 in 4 risk of having an affected child. In other cases the defect has been recognized. Phenylketonuria, which has an incidence of one in 15 000 births, affects the enzyme phenylalanine hydroxylase. Affected individuals are unable to convert phenylalanine to tyrosine and the resulting disturbance in metabolism leads to mental retardation. Fortunately it is possible to detect this deficiency at birth and by controlling the phenylalanine in the diet the disease can be prevented. For diseases in which the primary defect is known and for those in which it is not, it is clear that it will shortly be possible to offer ante-natal diagnosis using cloned DNA molecules as probes for blotting the chromosomal DNA of the unborn foetus. Such DNA can be readily obtained, at minimal risk, by the process of amniocentesis. This involves withdrawing a sample of amniotic fluid by inserting a syringe into the amniotic cavity at about the sixteenth week of pregnancy. The fluid contains foetal cells which can be cultured to provide enough material from which DNA can be extracted and a Southern blot analysis performed.

The first successes in this type of approach are with genetic diseases that affect the globin genes. This reflects the early advances made in studying the molecular biology of these genes. Globin genes from several mammalian species have now been cloned and shown to have two intervening sequences, a larger one corresponding to the sequence described by Jeffreys and Flavell (see Section 6.2.1) and a smaller one about a hundred base pairs long. Flavell and his coworkers have used Southern's blotting technique to analyse the organization of the human globin genes. The expression of the globin genes follows a developmentally regulated pattern. In the early embryo the haemoglobin molecule consists of two ζ-chains and two ε-chains (ζ_2, ε_2). After 3 months, the major foetal form of haemoglobin is HbF (α_2, γ_2). The γ chains are encoded by two non-allelic genes whose expression ceases at birth as HbF is replaced by HbA (α_2, β_2) and a low level of HbA_2 (α_2, δ_2). The cloned human β-globin cDNA [26] has been used as a hybridization probe for both the β-gene and the closely related δ-gene in chromosomal DNA. The two genes can be distinguished since, by increasing the stringency of the hybridization washes, it is possible to melt out those probe sequences hybridized to the δ gene. Experiments of this type showed that the β and δ genes were closely linked. The availability of cloned cDNAs to probe for the foetal genes (G_{γ} and A_{γ}) led to the establishment of a physical linkage map (see Fig. 6.16). Similarly a physical map has been established for the α-gene cluster.

There are many inherited defects that affect both the structural genes for the globins *per se*, and the various stages of their biosynthesis including their developmental regulation. Over 300 structural gene variants have been identified which have a variety of phenotypic consequences including cell sickling, the precipitation of globin chains, altered oxygen affinity and so on. In some cases when the base change that causes the amino-acid substitution lies within a restriction enzyme cleavage site, the mutation can be recognized by Southern blotting (e.g. HbOArab which removes an *Eco*RI site) [27]. Deletions can also be readily detected by Southern blotting analysis. Several types of α- or β-thalassaemia have been shown in this way to be a consequence of gene deletions. Thalassaemia is a condition in which the synthesis of α- or β-globin chains is reduced. The excess chains precipitate and cause haemolytic anaemia and enlargement of the spleen. Treatment involves regular blood transfusions and iron chelation therapy to remove iron generated by haem degradation. Many types of thalassaemia are deletions of the structural gene and can be readily detected by Southern blotting analysis. The extent of the deletions in a number of thalassaemias is illustrated in Fig. 6.16.

The mutations causing inherited abnormalities of globin synthesis are not limited to deletions and include aberrations affecting all levels of gene expression. The molecular basis of thalassaemia has been reviewed [28] and the reader is referred to this article for a detailed description of the different types of this disease. Mutations have been

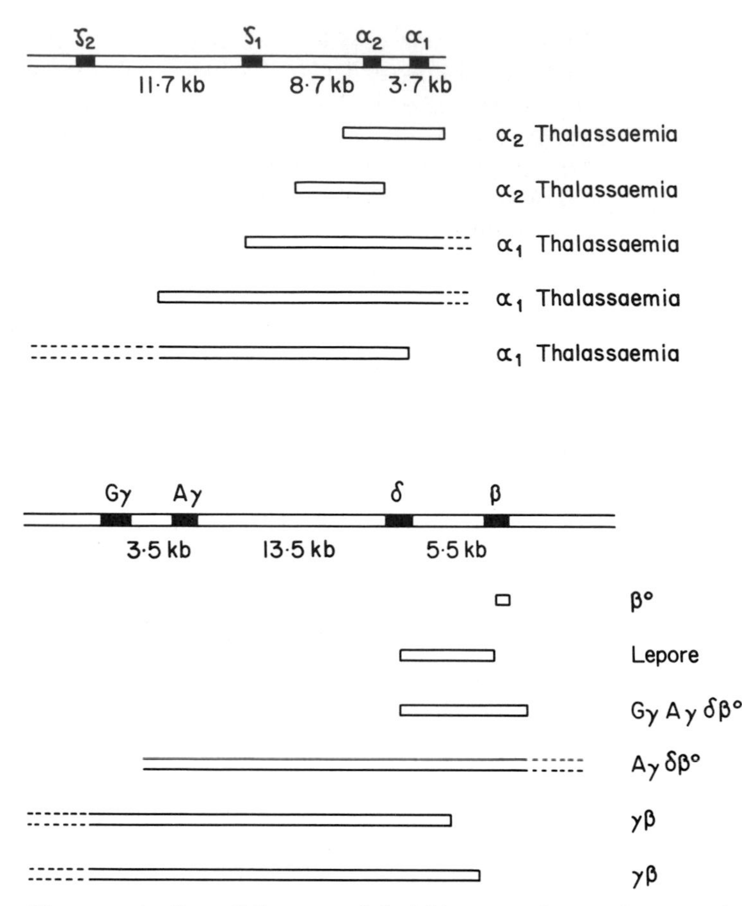

Figure 6.16 The organization of the α- and β-globin gene clusters in man showing sequences that are deleted in some thalassaemias.

found that affect the transcription of the β-globin gene. In one case there is a single nucleotide change in the 'TATA' box, a sequence found upstream of most eukaryotic genes and thought to be important in the initiation of transcription (see Chapter 8). In another case there is a single nucleotide substitution 87 nucleotides upstream from the transcription initiation site. There are also several examples of β^0-thalassaemia that result from nonsense mutations or frame shifts caused by insertions or deletions. Other β^0-thalassaemias are the result of mutations that affect splicing, the process whereby the intervening sequences are removed from the primary transcript during the maturation of RNA. Intervening sequences invariably have the dinucleotide sequence GT at their 5′ donor ends and the dinucleotide AG at their 3′ or acceptor site. Mutations affecting splicing can specifically alter these nucleotides, or they can create new donor or acceptor sites within either the coding sequences or within the intervening sequence itself. Different thalassaemias provide several examples of such splicing defects [29–31].

The ability to detect such mutations by direct analysis of DNA offers the approach to ante-natal diagnosis. Southern blotting experiments carried out on DNA derived from amniotic cells are highly suited for the analysis of gross deletions of chromosomal DNA. An alternative approach is needed, however, in order to detect mutations which do not lead to changes in the restriction cleavage pattern. The types of thalassaemia which are due to defects in splicing fall into this category. It should in many of these cases be possible to recognize mutations by virtue of their linkage to restriction site polymorphisms in the DNA flanking the gene. For example, the normal β-globin gene is contained within a 7.6 kb *Hpa*I fragment but that of HbS within a 13 kb fragment [32]. HbS is a defective haemoglobin which causes sickle-cell anaemia. It is due to a single base change that changes a glutamic acid codon (GAG) to a valine codon (GTG). The resulting haemoglobin tends to crystallize in the red cells which are consequently removed by the spleen. Similarly the β^0 gene is associated with Sardinian β^0-thalassaemia contained in a 9.3 kb *Bam*HI fragment, as compared to the 22 kb fragment which contains the normal gene [33]. The association of the mutation with a variant restriction site in flanking DNA is fortuitous. The two variants will, however, continue to be co-inherited unless they are separated by recombination. The chance of misdiagnosis due to recombination is negligible for restriction sites within a few thousand base pairs from the gene. The association of the sickle mutation with a 13 kb *Hpa*I fragment is found in the black population of the USA. This population originates largely from the west coast of Africa where affected blacks also show this linkage. In the black population in south and east Africa, however, both the sickle and normal gene are associated with the 7.6 kb *Hpa*I fragment. The restriction site polymorphism does, however, offer a useful diagnostic aid to affected families especially since a further polymorphism has been discovered of a *Hin*dIII site in the δ-gene linked to the HbS mutation (Fig. 6.17). These assays have now been superseded since restriction endonucleases have been discovered whose sites are affected by the specific base change of the HbS mutation. These are *Mst*II and *Dde*I which recognize the sequences CCTNAGG and CTNAG respectively.

If a convenient restriction site is not available and the sequence of the mutation is known, it is possible to synthesize an oligonucleotide that can be used as a probe to distinguish between wild type and mutant genes. We have seen an example of a similar approach in

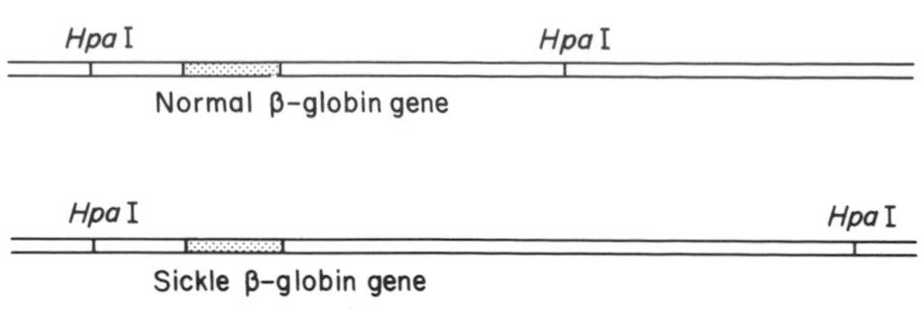

Figure 6.17 The association of a restriction fragment length polymorphism with the HbS mutation.

Chapter 3 where a synthetic oligonucleotide was used to distinguish between wild type and mutant Ml3 phages. This approach has been used to identify the dysfunctional anti-trypsin gene in the DNA of affected individuals. Anti-trypsin is a protein that inhibits the function of elastase and individuals lacking this protein suffer from diseases whereby the lung tissues are destroyed by elastase. The mutant gene has a single G to A change that results in the production of non-functional protein. This mutation can be detected in Southern blots using a 19 base oligomer complementary to the wild type gene as a probe. This approach can in principle be applied to diagnose any genetic disease provided that the precise nucleotide change is known.

Several genes have now been cloned which are known to be mutated in genetic diseases. The Lesch–Nyhan syndrome, which leads to a form of mental retardation, is an X-linked disorder characterized by the lack of hypoxanthine–guanine phosphoribosyl transferase (HGPRT). The gene for this enzyme has now been cloned and its detailed analysis could lead to a diagnostic screen. The absence of the urea-cycle enzyme arginosuccinate synthetase leads to another form of mental retardation, ammonia poisoning and early death. This gene has also been cloned and used as a probe to analyse the nucleic acids in cell cultures established from affected individuals. In many of these cases the defects seem to lie at the level of mRNA processing, and it is possible that 'Northern' blot analysis could be used diagnostically to detect these altered RNAs. The gene for phenylalanine hydroxylase has also been cloned and offers prospects for an ante-natal screen for phenylketonuria. In this last case it is possible that linked restriction endonuclease cleavage site polymorphisms could be used to identify the locus [34]. This approach should in future be applicable to a variety of genetic diseases. It is not necessary to identify the structural gene responsible for the disease, providing that a restriction site polymorphism can be identified which segregates within the dysfunctional gene. Indeed it should be possible to construct an entire map of the human chromosomes using such polymorphisms. In many genetic diseases the primary defect is not understood. Cystic fibrosis, muscular dystrophy and

Huntington's disease are examples of such diseases. Searches are being undertaken for restriction fragment length polymorphisms (of the kind described above for HbS) that provide a genetic marker linked to the locus responsible for the syndrome. A restriction fragment length polymorphism has been associated with Huntington's disease, which has allowed the locus to be mapped to chromosome 4 [35]. Huntington's disease is a progressive neurodegenerative disease which appears between the ages 30–50 years. The associated polymorphisms may provide a suitable ante-natal diagnosis if they prove to be sufficiently closely linked. Similarly two restriction fragment length polymorphisms have been mapped relative to the Duchenne muscular dystrophy locus on the short arm of the X chromosome [36]. These will assist in ordering further markers that are needed to establish an ante-natal screen.

References

1. Southern, E. M. (1979) Estimation of sizes of restriction fragments on agarose gels. *Analyt. Biochem.*, **100**, 319–24.
2. Boseley, P., Moss, T., Machler, M., Portmann, R. and Birnstiel, M. (1979) Sequence organisation of the spacer DNA in a ribosomal gene unit of *X. laevis*. *Cell*, **17**, 19–32.
3. Davis-Hershey, N., Conrad, S., Sodja, A. *et al.* (1977) The sequence arrangements of *Drosophila melanogaster* 5S DNA cloned in recombinant plasmids. *Cell*, **11**, 585–98.
4. Smith, D. F., McClelland, A., White, B. N., Addison, C. F. and Glover, D. M. (1981) The molecular cloning of a dispersed set of developmentally regulated genes which encode the major larval serum protein of *D. melanogaster*. *Cell*, **23**, 441–9.
5. Garapin, A. C., Cami, B., Roskam, W. *et al.* (1978) Electron microscopy and restriction enzyme mapping reveal additional intervening sequences in the chicken ovalbumin split gene. *Cell*, **14**, 629–39.
6. Wu, M., Holmes, D. S., Davidson, N., Cohn, R. H. and Kedes, L. H. (1976) The relative positions of sea urchin histone genes on the chimeric plasmids pSp2 and pSp17 as studied by electron microscopy. *Cell*, **9**, 163–70.
7. White, R. L. and Hogness, D. S. (1977) R loop mapping of the 18S and 28S sequences in the long and short repeating units of *Drosophila melanogaster* rDNA. *Cell*, **10**, 177–92.
8. Brack, C., Hirama, M., Lenhard-Schuller, R. and Tonegawa, S. (1978) A complete immunoglobulin gene is created by somatic recombination. *Cell*, **15**, 1–14.
9. Marcu, K. B. (1982) Immunoglobulin heavy chain constant region genes. *Cell*, **29**, 719–21.
10. Baltimore, D. (1981) Somatic mutation gains its place among the generators of diversity. *Cell*, **26**, 295–6.
11. Berk, A. J. and Sharp, P. A. (1977) Sizing and mapping of early

adenovirus mRNAs by gel electrophoresis of S1 endonuclease digested hybrids. *Cell*, **12**, 721–632.
12. Birnstiel, M. L., Sells, B. M. and Purdom, I. F. (1972) Kinetic complexity of RNA molecules. *J. Mol. Biol.*, **63**, 21–39.
13. Southern, E. M. (1975) Detection of specific sequences among DNA fragments separated by gel electrophoresis. *J. Mol. Biol.*, **98**, 503–17.
14. Laskey, R. A. and Mills, A. D. (1977) Enhanced autoradiographic detection of ^{32}P and ^{125}I using intensifying screens and hypersensitized film. *FEBS Lett.*, **82**, 314–6.
15. Rigby, P. W. J., Dieckmann, M., Rhodes, C. and Berg, P. (1977) Labelling deoxyribonucleic acid to high specific activity *in vitro* by nick translation with DNA polymerase 1′. *J. Mol. Biol.*, **113**, 237–51.
16. Jeffreys, A. J. and Flavell, R. A. (1977) The rabbit β-globin gene contains a large insert in the coding sequence. *Cell*, **12**, 1097–108.
17. Maniatis, T., Kee, S. E., Efstratiadis, A. and Kafatos, F. (1976) Amplification and characterisation of a β-globin gene synthesised *in vitro*. *Cell*, **8**, 163–82.
18. Alwine, J. C., Kemp, D. J. and Stark, G. R. (1977) Method for detection of specific RNAs in agarose gels by transfer to diazobenzyloxymethyl paper and hybridisation with DNA probes. *Proc. Natn Acad. Sci. USA*, **74**, 5350–4.
19. Roop, D. R., Nordstrom, J. L., Tsai, S. Y., Tsai, M.-J. and O'Malley, B. W. (1978) Transcription of structural and intervening sequences in the ovalbumin gene and identification of potential ovalbumin mRNA precursors. *Cell*, **15**, 671–85.
20. Strobel, E., Dunsmuir, P. and Rubin, G. M. (1979) Polymorphisms in the chromosomal locations of elements of the 412 copia and 297 dispersed repeated gene families in *Drosophila*. *Cell*, **17**, 429–39.
21. Harper, M. E., Ullrich, A. and Sanders, G. F. (1981) Localisation of the human insulin gene to the distal end of the short arm of chromosome 11. *Proc. Natn Acad. Sci. USA*, **78**, 4458–60.
22. Trent, J. M., Olson, S. and Lawn, R. M. (1982) Chromosomal localisation of human leukocyte, fibroblast and immune interferon genes by means of *in situ* hybridisation. *Proc. Natn Acad. Sci. USA*, **79**, 7809–18.
23. Jeffreys, A., Craig, I. N. and Francke, U. (1979) Localisation of the $^{G}\gamma$-, $^{A}\gamma$-, δ- and β-globin genes on the short arm of human chromosome 11. *Nature*, **281**, 606–8.
24. Owerbach, D., Bell, G., Rutter, W. J. and Showa, T. B. (1980) The insulin gene is located on chromosome 11 in humans. *Nature*, **286**, 82–4.
25. Davies, K. E., Young, B. D., Elles, R. G., Hill, M. E. and Williamson, R. (1981) Cloning of a representative genomic library of the human X chromosome after sorting by flow cytometry. *Nature*, **293**, 374–6.
26. Konkel, D. A., Tilghman, S. M. and Leder, P. (1978) The sequence of the chromosomal mouse β globin major gene: homologues in capping, splicing and poly(A) sites. *Cell*, **15**, 1125–32.
27. Flavell, R. A., Kooter, J. M., de Boer, E., Little, P. F. R. and Williamson, R. (1978) Analysis of the β-δ-globin gene loci in normal and Hb lepore DNA: direct determination of gene linkage and intergene distance. *Cell*, **15**, 25–42.

28. Weatherall, D. J. and Clegg, J. B. (1982) Thalassemia Revisited. *Cell*, **29**, 7–9.
29. Busslinger, M., Moschonas, N. and Flavell, R. A. (1981) β^+ Thalassemia: apparent splicing results from a single point mutation in an intron. *Cell*, **27**, 289–98.
30. Orkin, S. H., Goff, S. C. and Hechtman, R. L. (1981) Mutation in an intervening sequence splice junction in man. *Proc. Natn Acad. Sci. USA*, **78**, 5041–5.
31. Treisman, R. A., Proudfoot, N. J., Shandler, M. and Maniatis, T. (1982) A single base change at a splice site β^0 thalassemic gene causes abnormal RNA splicing. *Cell*, **29**, 903–11.
32. Kan, Y. W. and Dozy, A. M. (1978) Antenatal diagnosis of sickle cell anaemia by DNA analysis of amniotic fluid cells. *Lancet*, **ii**, 910–2.
33. Kan, Y. W., Lee, K. Y., Furbetta, M., Angus, A. and Cao, A. (1980) Polymorphism of DNA sequence in the β-globin gene region – application to prenatal diagnosis of β^0 thalassemia in Sardinia. *New Engl. J. Med.*, **302**, 185–8.
34. Woo, S. L., Lidsky, A. S., Guttler, F., Chandra, T. and Robson, J. H. (1983) Cloned human phenylalaninehydroxylase gene allows prenatal diagnosis and carrier detection of classical phenylketonuria. *Nature*, **306**, 151–5.
35. Gusella, J. F., Wexler, N. S., Conneally, M. *et al.* (1983) A polymorphic DNA marker genetically linked to Huntington's disease. *Nature*, **306**, 234–8.
36. Davies, K. E., Pearson, P. L., Harper, P. S. *et al.* (1983) Linkage analysis of two cloned DNA sequences flanking the Duchenne muscular dystrophy locus on the short arm of the human X chromosome. *Nuc. Acids Res.*, **11**, 2303–12.

7 Gene cloning in fungi and plants

7.1 *Saccharomyces cerevisiae*

Yeast offers a number of advantages as a unicellular eukaryotic organism. It can be propagated in either the haploid or diploid state. It is capable of sporulation to form spore tetrads, from which individual haploid spores can be dissected and individually propagated. These natural advantages of the organism have led to its extensive genetic characterization and have played an important part in stimulating the development of gene cloning systems. DNA can be introduced into yeast cells which have been treated with a mixture of β-gluconases from snail gut extract to digest polysaccharide in the cell wall. The spheroplasts, so produced, can be maintained in isotonic solutions of sorbitol and DNA uptake can be promoted by the addition of polyethylene glycol. The cell wall regenerates if the treated spheroplasts are plated in 3% agar.

7.1.1 Integrating vectors

One of the first demonstrations that yeast protoplasts could be transformed with DNA was the work of Hinnen *et al.* [1], who were able to select yeast cells which had taken up and expressed the yeast *leu*2 gene carried on ColEl (see Chapter 1). The transformants contain DNA that has integrated into homologous sequences on the yeast chromosome. (In the original work of Hinnen *et al.* it was thought that integration could also occur at non-homologous sites. It is now known that the plasmid which was used in these experiments contained a sequence that is repeated within the yeast genome. The integrations that were thought to be at non-homologous sites were in fact into these repetitive sequences at their other locations. Integration can result in replacement of the yeast chromosomal gene by the one carried on the plasmid without the vector sequences being incorporated into the yeast chromosome. These transformants must arise either by a double cross-over event or by gene conversion. In the majority of transformation events, the vector sequences also integrate into the yeast chromosome (Fig. 7.1) and can be shown to segregate in a Mendelian manner.

The duplicated structure produced by the integration of the cloned yeast gene and the vector is not stable. Struhl *et al.* [2] showed, in the case of a *his* transformant generated as described above, that after 15 generations approximately 1% of colonies were *his*⁻ segregants, and these cells had completely lost the transforming DNA by an intrachromosomal crossover. This observation was used by Scherer and Davis [3] as the basis of a means for exchanging a chromosomal segment of yeast DNA for the homologous segment which contains an insertion or deletion introduced *in vitro*. The principle of this

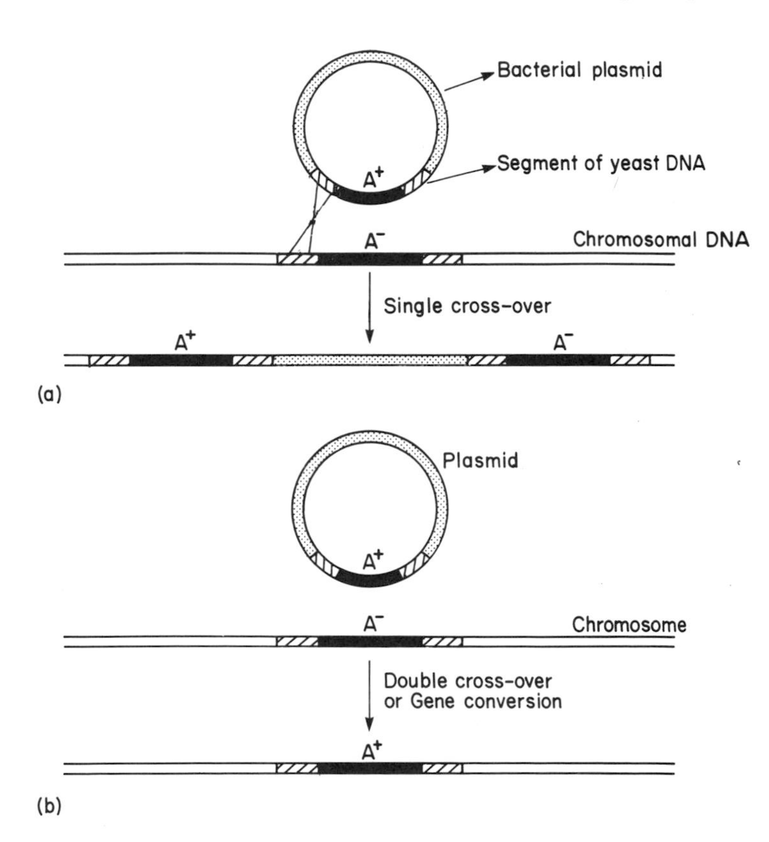

Figure 7.1 Integration of cloned DNA into the yeast chromosome during transformation.

method, which is now known as *transplacement*, is illustrated in Fig. 7.2. In this example the yeast *his*3 gene is carried on a bacterial plasmid together with the yeast *ura*3 gene. The *ura*3 gene is used only as a selective marker and since the recipient yeast strain has a *ura*3 deletion, homologous recombination occurs between the plasmid and the *his*3 gene on the yeast chromosome. The *his*3 gene on the plasmid has a 150 base pair deletion generated by *Hin*dIII cleavage and religation. Following the selection of transformants, the selective pressure is removed and *ura*$^-$ segregants are recovered. These have lost the *ura*$^+$ gene and linked bacterial sequences by recombination between the pre-existing sequences at or around the *his*$^+$ locus and the newly introduced homologous sequences. Two classes of segregants are identified, one of which is indistinguishable from the non-transformed strain, and the other in which the *his*3$^+$ locus has been replaced by the *his*3 gene containing the 150 base pair deletion. This procedure is extremely useful in introducing a mutation into a given strain to produce a counterpart that is otherwise isogenic.

Integrative plasmids have also been used to mutagenize chromosomal loci. This is known as gene *disruption*. It has been used to obtain

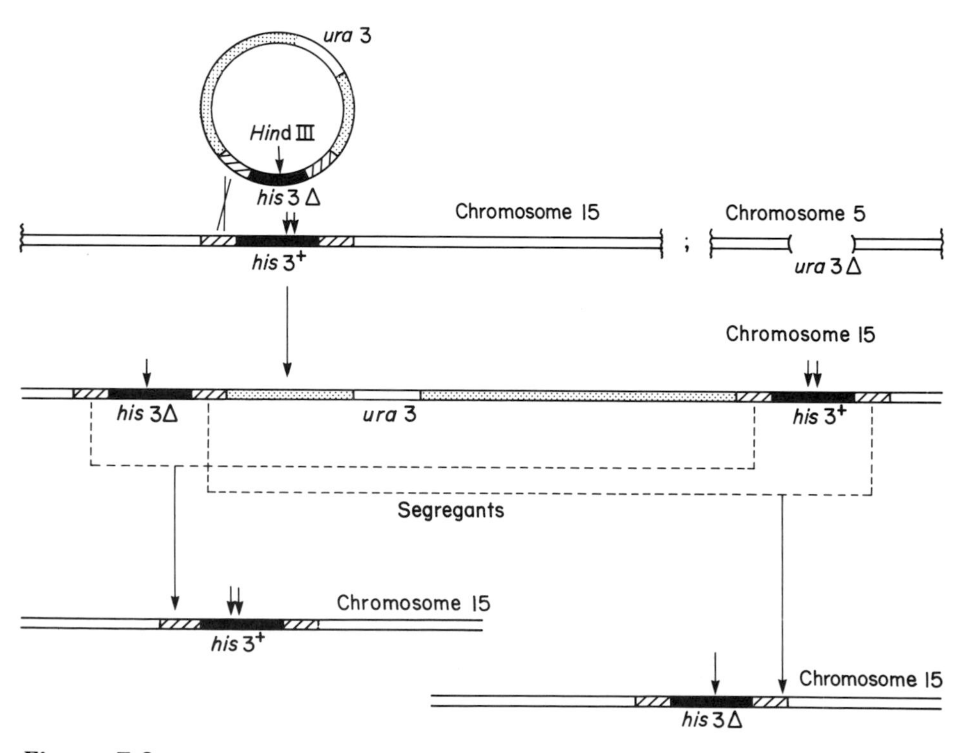

Figure 7.2 Transplacement of a yeast chromosomal segment with a *his*3 gene having a deletion.

definitive identification of cloned genes. The principle of the method is shown in Fig. 7.3 for the cloned gene for actin. This was isolated from a library of cloned yeast genes by its homology to the cloned actin genes from other eukaryotes. This plasmid was introduced into a diploid strain of yeast where it integrated into one of the wild type alleles of the resident actin genes. This results in a mutation that segregates as a recessive lethal closely linked to the selectable marker gene (in this case *ura*3) on the integrated plasmid [4]. A similar procedure has been used to identify an essential function associated with the β-tubulin gene [5].

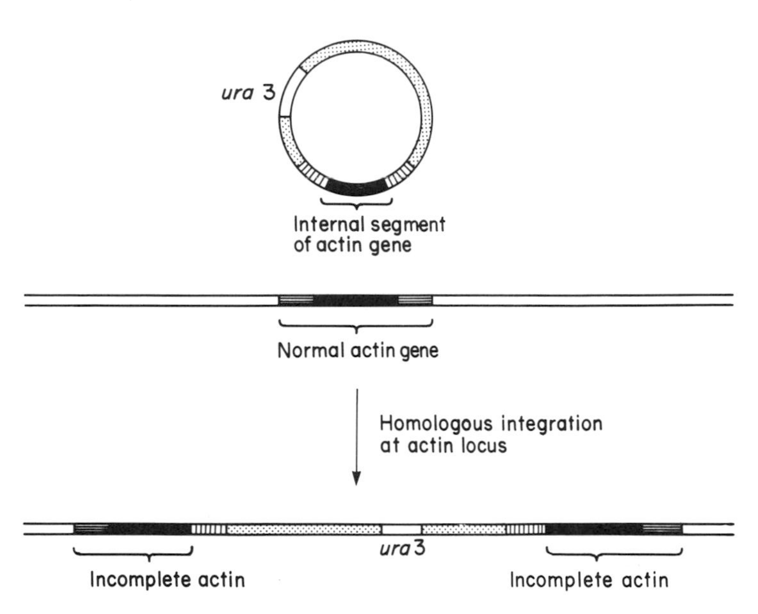

Figure 7.3 Lethal disruption of the yeast actin gene by integrative DNA transformation.

Integration is a rare event when circular plasmid DNA carrying a yeast gene is introduced into cells. If a double stranded break is introduced into the plasmid by restriction endonuclease digestion, however, the integration frequency increases by about 100-fold to give one transformant in about 10^4 cells [6]. If a plasmid contains two yeast genes, '*A*' and '*B*', one can preferentially direct the integration of the plasmid into one or other of the corresponding chromosomal genes by cleaving the plasmid in one of the two genes. If the plasmid is cleaved by a restriction endonuclease that cuts in gene '*A*' then integration will occur preferentially into chromosomal gene '*A*' and *vice versa.* A plasmid cut twice within a yeast fragment also transforms at high frequencies. The gap that is generated in this way between the two restriction sites is repaired during the integration event. Vector sequences do not have to be present and it is possible to introduce a linear yeast DNA fragment directly into the chromosome because of the highly recombinogenic nature of free DNA ends in yeast [7]. An example of this approach used for gene disruption is shown if Fig. 7.4. A selectable gene, '*A*', is introduced into the gene of interest, '*B*', and a linear restriction fragment containing this gene is then prepared. This is introduced into A^- yeast spheroplasts and amongst the transformants are cells which are A^+ and B^-.

In most cases, transformation with the integrative vectors introduces bacterial vector sequences into the yeast chromosome. These bacterial vector sequences have been used as a means of *retrieving* specific yeast genes from the site of integration. DNA from the transformed yeast strain is cleaved with a restriction endonuclease that does not cleave within the bacterial vector. The resulting fragments are cyclized using DNA ligase and used to

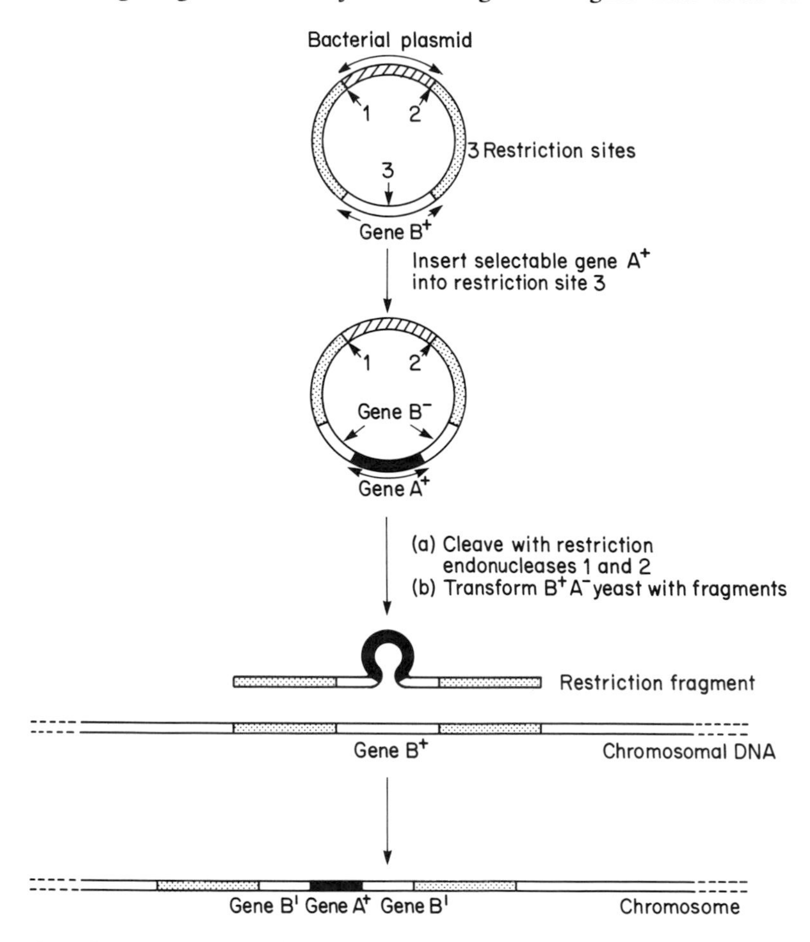

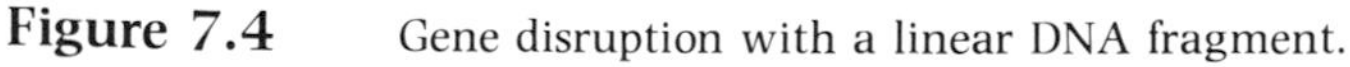

Figure 7.4 Gene disruption with a linear DNA fragment.

transform *E. coli*. The bacterial transformants, selected by the marker carried on the bacterial plasmid, contain a cloned segment of yeast DNA from the original site of integration. The technique is often used if one has cloned a wild type allele of a yeast gene and one wishes to examine mutations in that gene in different yeast strains. The cloned gene is introduced into the mutant strain such that following integration there will be tandem copies of the gene separated by the *E. coli* vector sequences. The plasmid is constructed so that the yeast DNA is flanked by two unique restriction sites. Cleavage of the chromosomal DNA with one of these enzymes will therefore generate a DNA fragment containing the mutant gene. Conversely cleavage with the other enzyme will generate the wild type gene (Fig. 7.5).

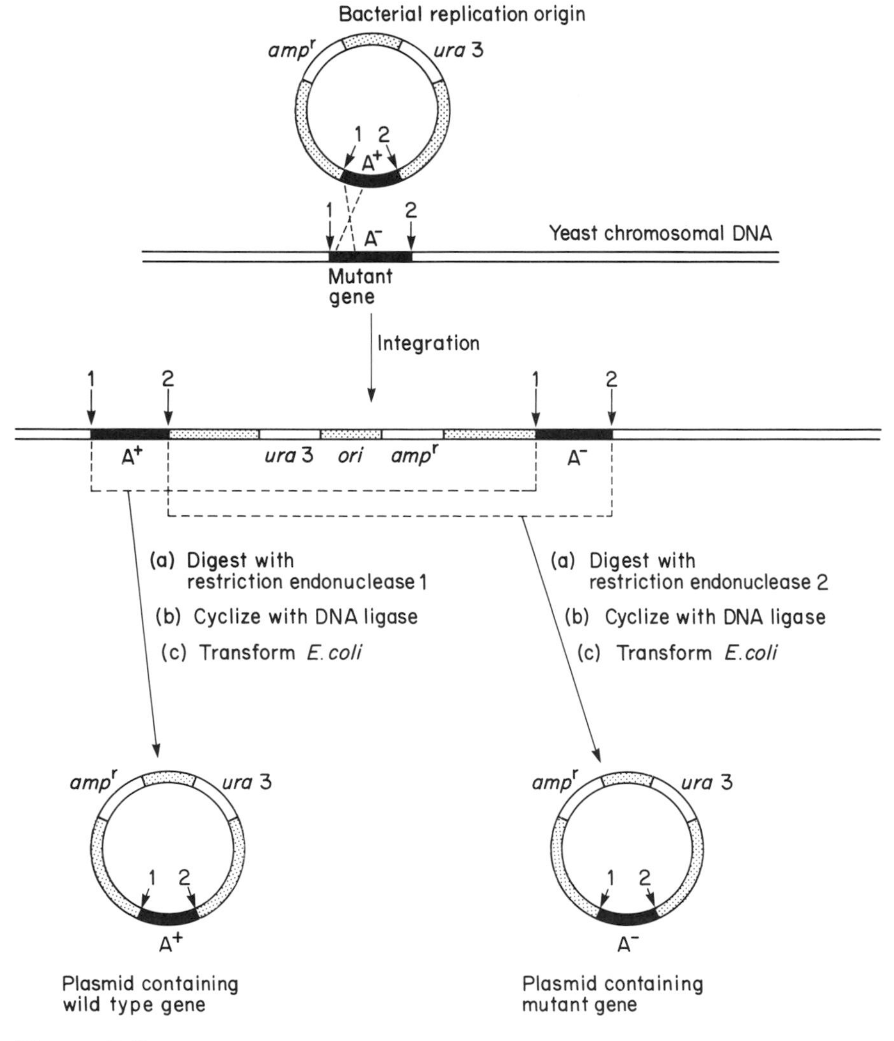

Figure 7.5 Retrieving a gene from the yeast chromosome.

7.1.2 Autonomously replicating vectors

The transformation frequency of yeast cells by recombinant plasmids containing such markers as *his*3, *leu*2 and *ura*3 is about 10^{-7} transformants per viable cell. On the other hand segments of yeast DNA linked to the *trp*1 and *arg*4 genes have been demonstrated to transform at a much higher frequency (about 10^{-4}) [8, 9]. In these cases the transforming molecules replicate autonomously within the cells. The DNA segments responsible for this property have been termed *ars* (autonomously replicating sequence) elements. DNA segments which have the *ars* property have been isolated from a wide range of eukaryotic organisms, including *Dictyostelium discoideum*, *Caenhaorabditis elegans*, *Drosophila melanogaster* and *Zea mays* [10]. They all share a consensus sequence AAA(C/T)ATAAA which lies within an AT-rich region. It is not known whether these sequences act as replication origins in the chromosomal DNA in their species of origin. The transformants obtained with *ars* vectors are unstable and their plasmids are rapidly eliminated if selective pressure is taken away. The instability of *ars* elements is due to irregular segregation at mitosis. The plasmids are, however, stabilized when joined to centromeric sequences. The centromeric sequences from chromosome 3 were isolated by 'walking' from the *leu*2 gene, known to map to the left of the centromere, to the *pgk* (phosphoglycerokinase) gene to the right. One of these cloned DNA fragments confers stability on a plasmid containing an *ars* element, making it behave as a mini-chromosome which segregates in a Mendelian manner [11]. A 6–10 kb DNA segment, *cen*, is needed for DNA molecules to segregate correctly in both meiosis and mitosis. Smaller sub-cloned regions are capable of stabilizing plasmids in mitosis, but they will not direct proper segregation in meiosis. *Cen* elements are useful components of yeast vector plasmids since they confer stability and thereby control the gene dosage.

Telomeric sequences of chromosomes have also been functionally identified in yeast. These were initially isolated from the ribosomal DNA of *Tetrahymena*. In the macro-nucleus of this organism the rDNA is maintained extrachromosomally as a linear palindromic molecule in which the ends are covalently joined to give hairpin structures. When the terminal fragments of *Tetrahymena* rDNA molecules were ligated to the ends of a linear molecule comprised of a yeast marker gene and an *ars* element, the construct was found to replicate in yeast as a linear molecule [12].

Another method of achieving efficient transformation of yeast cells uses a set of vectors derived from the endogenous 2 μm yeast plasmid [13]. The plasmid is present in many strains of yeast in 50–100 copies per cell. Cloning vectors have been constructed which are *in vitro* recombinants between this plasmid and bacterial cloning vectors such as pMB9, pBR322 and pBR325. Such vectors will replicate in either *E. coli* or yeast cells. Similar vectors have been developed for the fission yeast, *Schizosaccharomyces pombe* [14]. These are based either on *ars* sequences from *S. pombe* or on the *S. cerevisiae* 2 μm plasmid. They are suitable for shuttling DNA between these two species. This approach has been used, for example, to isolate an *S. cerevisiae* gene that will complement an *S. pombe* cell division cycle mutant [15].

Genes can also be retrieved from the yeast chromosomes using vectors that have yeast replication origins. The method utilizes a 'gapped' molecule which is highly recombinogenic and so integrates at a high frequency. The wild type gene is first removed from the 'retriever plasmid' by cleavage with restriction endonucleases in such a way that the sequences which flank the gene in the chromosome are left behind (see Fig. 7.6). The remainder of the molecule is introduced into yeast cells, where in order for it to replicate it must form a circle. This is achieved by recombination

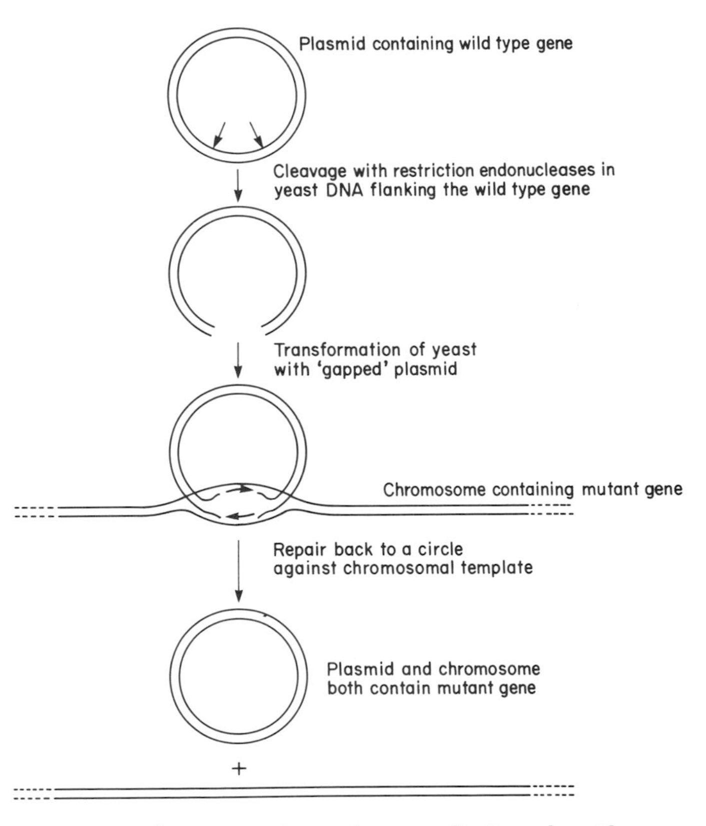

Figure 7.6 Retrieving a chromosomal gene into a replicating plasmid.

between the sequences that flank the wild type gene on the plasmid and the corresponding chromosomal sequences in the mutant strain. The missing segment of the gene carried on the plasmid is replaced by a copy of the mutant chromosomal gene during DNA replication. Selection for a marker gene on the plasmid allows the isolation of colonies carrying circular plasmid molecules.

7.1.3 Expression of foreign DNA in yeast

As a eukaryotic microorganism, yeast initially seemed a very attractive system for the expression of cloned genes. In some cases these hopes have been realized whereas in others, experience has moderated the initial enthusiasm. The gene for β-lactamase is an example of a prokaryotic gene that is expressed in yeast. If yeast cells carrying a plasmid with the β-lactamase gene are plated with *E. coli* in the presence of ampicillin, the yeast colonies permit a halo of bacterial growth around themselves. In *E. coli*, the β-lactamase pre-protein has a signal peptide that is cleaved off during transport into the periplasmic space. The inactive pre-protein is apparently converted to mature β-lactamase within the yeast cell [16]. The bacterial gene for chloramphenicol transacetylase is also expressed in yeast [17]. Neither of these antibiotics, however, are suitable for a selective scheme in yeast. On the other hand the 2-deoxystreptamine antibiotic, G418, prevents the growth of many yeasts by inhibiting protein synthesis. This antibiotic is inactivated by an aminoglycoside 3′ phosphotransferase encoded by the prokaryotic transposon Tn5. This confers bacteria with resistance to neomycin and kanamycin. This gene is expressed in yeast and confers resistance to G418 [18]. It may therefore offer a useful selectable marker for cloning in yeast and also in animal cells (see Section 8.2.3).

The initial studies on the expression of higher eukaryotic genes in yeast were disappointing. Several studies indicated that intervening sequences cannot be removed from the transcripts of foreign genes in yeast. This is apparently due to the need for a specific octanucleotide sequence in the intervening sequence. This sequence is found in yeast genes about 20–25 nucleotides upstream from the 3′ splice site [19]. It is therefore more appropriate to use cDNA clones in order to achieve expression in yeast, as was discussed in Chapter 1. Considerable success has now been achieved with expression vectors that utilize yeast promoters. In some cases this has been by good fortune and in others by good management. One case of fortuitous expression was from a yeast promoter to the 3′ end of the *his*3 gene [20]. In this particular experiment the herpes simplex virus thymidine kinase (*tk*) gene was expressed. Yeast cells expressing the herpes *tk* gene can be grown in the presence of amethopterin and sulphamilamide (inhibitors of dTMP synthesis) if thymidine is supplied in the medium. This also has some potential as a selectable marker and it offers a system whereby yeast promoters could be specifically selected. Yeast promoters have also been deliberately introduced into vectors. A segment of DNA containing the promoter of the yeast alcohol dehydrogenase gene has been successfully used to permit high levels of expression of human leukocyte interferon D [21] and of hepatitis B virus surface antigen [22]. In the latter case the antigen is synthesized in yeast in the form of particles or aggregates with an appearance similar to empty virus particles and which are capable of inducing antibodies in animals. A similarly effective construct has the hepatitis B surface antigen gene under the control of the promoter of a yeast acid phosphatase gene [23]. Some genes are poorly expressed in yeast. This could be for a variety of reasons. It may be that in order to get efficient translation, the sequence context in the vicinity of the initiator ATG codon of the foreign gene needs to resemble that of a yeast gene. Alternatively codon usage may be important. The cDNA encoding rat growth hormone, for example, does not direct protein synthesis in yeast. In this gene, leucine is encoded at two positions by two codons which are very rarely used in yeast. The availability of tRNAs for these codons may then become a limiting factor for efficient expression. It is, however, not clear whether this is an important factor and the major limitation may well be factors governing the stability of foreign mRNA.

7.2 Filamentous fungi

Vector systems for genetically well characterized species of filamentous fungi are rapidly being developed. For *Neurospora crassa*, for example, a system similar to that used for yeast has been developed to transform freshly germinated conidia [24]. This particular method used the *N. crassa* gene for dehydroquinase (*qa*.2) which had been previously cloned in pBR322 by complementation of *E. coli ars*D mutants. One report suggests that the *N. crassa* DNA contained in this plasmid is capable of some autonomous replication when reintroduced as a plasmid into *N. crassa* [25]. Another laboratory has incorporated segments of the *Neurospora* mitochondrial genome into a similar plasmid. One of these fragments increases the transformation frequency 5–10-fold. The mitochondrial segment appears, however, to undergo deletion upon passage and the deleted derivative plasmid still appears to replicate [26]. A similar protocol has also been devised to introduce DNA into the protoplasts of *Aspergillus nidulans* in the presence of polyethylene glycol and calcium chloride [27]. This system was established using the cloned acetamidase gene to transform a strain with a deletion in the chromosomal acetamidase gene. The transforming DNA becomes integrated into the *Aspergillus* chromosome. Undoubtedly the experience gained with yeast will accelerate the development of cloning systems for the filamentous fungi.

7.3 Cloning DNA in plant cells – Ti plasmid

Success in cloning in bacteria, yeast and mammalian cell systems has resulted from the development of plasmid and viral vectors and it is toward such vectors that a number of groups are now looking to develop methods for the cloning and expression of foreign genes in plant cells. The DNA viruses of plants are good candidates for vectors. They are of two types: the Caulimoviruses which contain double-stranded DNA and the Gemini viruses which contain single-stranded DNA. Although each group has potential for development as cloning vehicles, this has as yet not been fully exploited. Undoubtedly this will happen over the next few years.

Viruses are not the only infectious agents with potential as vectors for plants. Crown Gall disease is caused by the infection of plants with certain strains of the soil-borne bacterium, *Agrobacterium tumefaciens*. This infection results in large tumour-like growths of plant tissue, usually on the stem of the plant just above the soil line. The causative agent is a bacterial plasmid, part of which – the 'T DNA' – integrates into the chromosomal DNA of the plant. It is therefore a naturally occurring vector for the transfer of foreign genes and it now looks as though it will find general application as a vector to introduce *in vitro* recombinant DNA into plants. Crown Galls can be experimentally induced in a variety of gymnosperms and dicotyledonous angiosperm plants by wounding and infecting with virulent strains of *Agrobacterium* [28]. The role of a plasmid was first suspected from the discovery that it was possible to transfer the property of virulence from a virulent strain to a non-virulent one if the two strains were innoculated into the same plant, although the tumour inducing (Ti) plasmid was not physically isolated until the early 1970s. Virulent strains which are cured of this plasmid become non-virulent [29]. and conversely strains that acquire the plasmid also acquire virulence. Virulence can be acquired in this way by other bacterial species, such as *Rhizobium trifolii* which exists symbiotically in the root nodules of legumes.

The infected plant cells acquire a segment of plasmid DNA that enables them to produce one or more unusual amino-acid derivatives known as opines. The enzymes for the biosynthesis of the opines are encoded by the plasmid. Opines are not normally found in plant tissue and are not used by the plant, but they provide a carbon and nitrogen source for the proliferation of the infecting bacteria. The enzymes for opine degradation are also encoded by the Ti plasmid, thus enabling the growth of bacteria carrying the plasmid in the plant tumour tissue. This symbiotic relationship of the plasmid and the bacterium is illustrated in Fig. 7.7.

The opines are formed by the reductive condensation of pyruvate

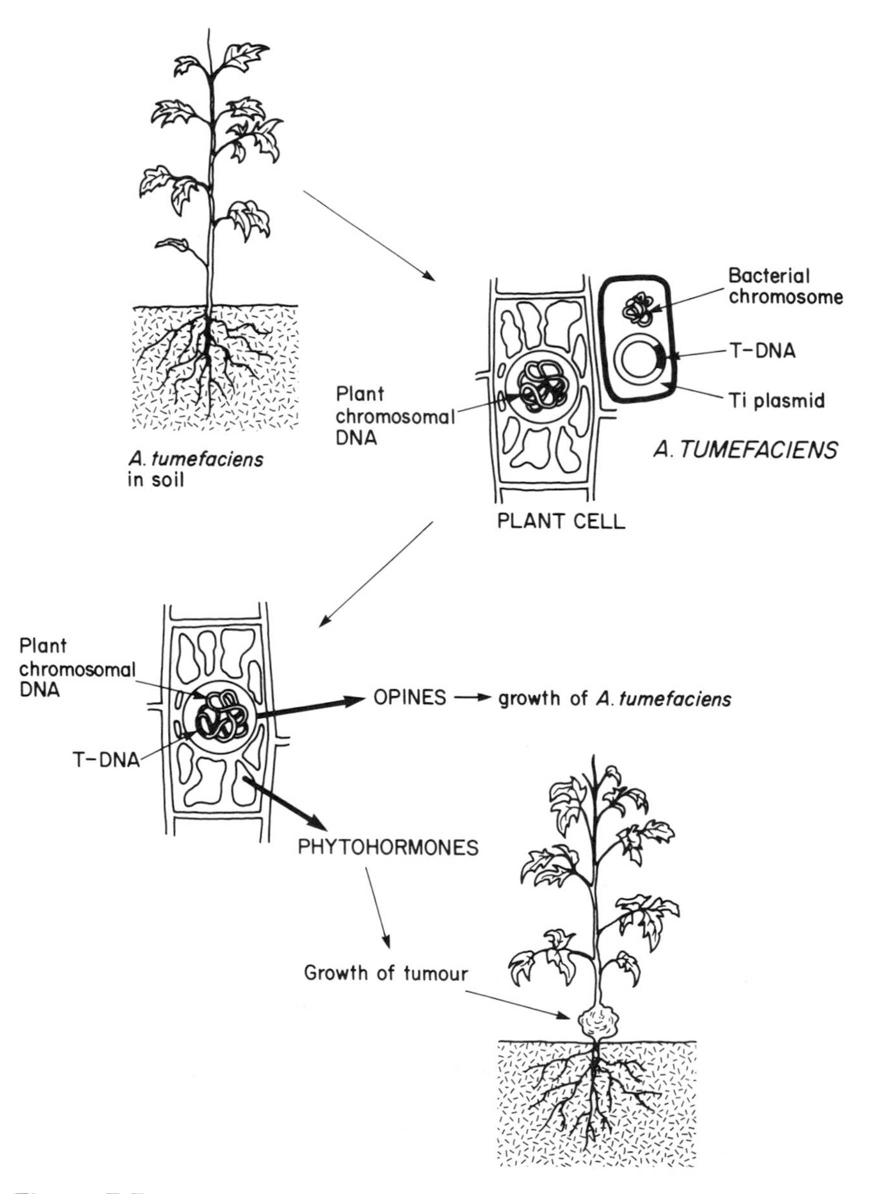

Figure 7.7 The induction of plant tumours by strains of *Agrobacterium* carrying the Ti plasmid.

or α-ketoglutarate with a basic amino-acid and the particular opine which is produced is a characteristic of the type of Ti plasmid. The two groups of plasmids which have been most extensively studied are those which specify the production of the opines, octopine and nopaline. There is considerable sequence diversity between these plasmids, even within a group specifying a particular opine. The T-DNA segments of the different groups do, however, show considerable sequence homology.

7.3.1 Genetic and physical maps of Ti plasmids

The Ti plasmids are very large ranging from 150–200 kb. They have been mapped using a combination of transposon mutagenesis and physical mapping techniques. The random insertion of transposons into the plasmid alters phenotypes specified by the plasmid. The position of the mutation is also 'flagged' so that it can be localized using either electron microscopy or 'Southern blotting' techniques (as described in Chapter 6). Furthermore, a restriction fragment carrying the transposon can also be recognized since it specifies markers, usually for drug resistance, carried on the transposon. The results of such mapping studies are shown for an octopine type plasmid and a nopaline type plasmid in Fig. 7.8. The genes which determine oncogenicity (*onc*) are distributed over wide segments of the plasmids and can also be found as chromosomal markers in the host bacterium [30]. The genes responsible for the biosynthesis of opines (octopine, *ocs*; nopaline, *nos*) as well as genes specifying their degradation (*occ*; *ncc*) are found at well defined regions. The system for the utilization of opines by the bacterial cell requires their active transport into the cell and subsequent cleavage to give the parent amino-acid and an α-keto acid, both being plasmid encoded functions. Both octopine and nopaline plasmids also encode one or more unidentified enzymes in the pathway for the degradation of arginine (*arc*), the parent amino-acid for these opines. Mutants unable to catabolize octopine have been selected as cells resistant to toxic analogues of octopine. These mutants synthesize normal levels of octopine showing that the biosynthetic and degradative processes are independent [31]. In addition, the plasmid carries genes (*tra*), which promote plasmid transfer from one bacterial strain to another; genes which confer the host with sensitivity to the antibiotic agrocin, produced by certain avirulent strains of *Agrobacterium*; genes which affect the host range of plants that *Agrobacterium* is able to infect; genes (*ape*) which give the ability to exclude the bacteriophage API; and genes which confer incompatibility (*Inc*) on the plasmid and which therefore govern its ability to coexist with Ti plasmids of other

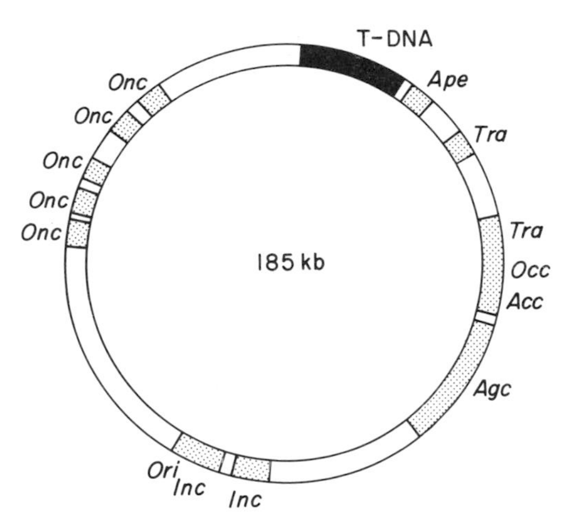

The octopine plasmid Ti-B806

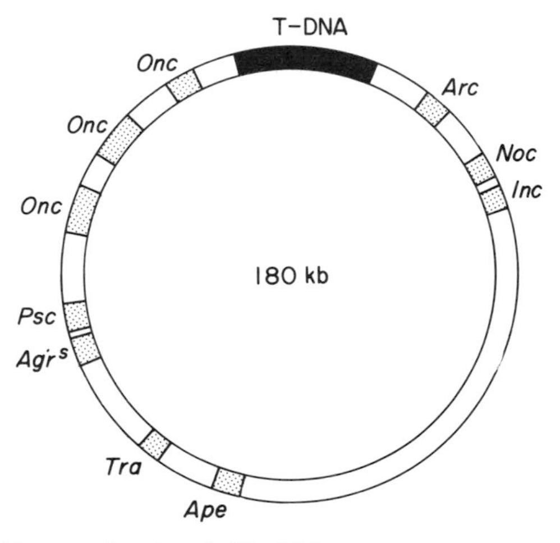

The nopaline plasmid Ti-C58

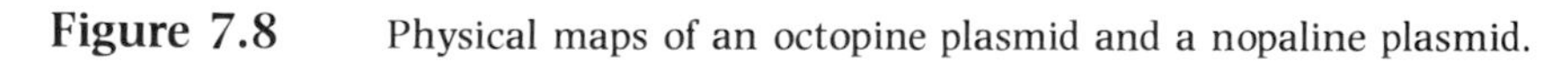

Figure 7.8 Physical maps of an octopine plasmid and a nopaline plasmid.

types. The map positions of these various functions are shown in Fig. 7.8.

7.3.2 T-DNA

During the process of plant cell transformation, plasmid DNA is transferred from the bacterium to the plant cell. The mechanism of this process is not understood. Southern blotting analysis has shown that a segment corresponding to roughly 10% of the Ti plasmid genome becomes stably associated with the plant cell. This is the segment known as T-DNA (Fig. 7.7). Subcellular fractionation studies have shown that T-DNA is incorporated into the nuclei of tumour cell DNA [32, 33]. Restriction analysis by Southern blotting has shown that the T-DNA is integrated at different chromosomal sites in different tumour cell lines. A number of such tumour cell lines have been analysed, allowing some generalizations to be made in terms of the segment of the plasmid which constitutes the T-DNA. The T-DNA complement differs in different tumour lines, but always contains a 'minimum sequence' known as 'core' T-DNA that is colinear with the Ti plasmid. The points of attachment of the T-DNA to the plant DNA are not absolute, although in a group of octopine tumours that have been studied, the left plasmid junction seems similar whereas the right junction is not at a unique position and the amount of DNA from this right end is extremely variable [34, 35]. In some tumours there are additional non-contiguous segments of Ti plasmid DNA.

As mentioned above a secondary advantage of transposon mutagenesis applied to Ti plasmid in its prokaryotic environment is that it tags T-DNA with an easily selectable marker. Ultimately this permits the recovery of T-DNA from the DNA of transformed plant cells. One example of this approach is the work of Holsters *et al.* [36] who made a cosmid library from the DNA of plant cells transformed with T-DNA containing the transposon Tn7. They were able to select bacteria transformed by cosmids specifying spectinomycin resistance (encoded by Tn7), and so identify clones of the T-DNA and flanking plant sequences. The Tn7 transposon was furthermore still capable of normal transposition showing that it had not undergone any drastic rearrangements whilst residing on the plant cell chromosome. This approach has been used to isolate the junctions between the plant cell DNA and the T-DNA for nucleic acid sequencing. The general picture which emerges from several studies is that the T-region in both nopaline and octopine plasmids is flanked by a direct repeat of 25 base pairs [37, 38, 39]. The borders of the T-DNA in the plant genome are frequently located either within or nearby this sequence. The sequencing studies also confirm earlier observations that tandemly arrayed T-DNA sequences are common in the plant tumours. The tandem units are separated by short segments of rearranged plant DNA sequences.

The biosynthesis of opines in transformed plant cells is consistent with the genetic localization of the opine biosynthesis genes within the T-DNA region of the plasmid (see Fig. 7.7). The identification of the restriction fragments containing the T-DNA region has allowed this region to be cloned and subjected to extensive transposon mutagenesis. The mutated fragment is then transferred back onto the Ti-plasmid by means of a double recombination event *in vivo*. This has permitted a fine structure genetic analysis upon the T-DNA region of two octopine type plasmids [40, 41]. These studies have identified loci which control tumour morphology. The loci fall into three categories: *tml* which causes undifferentiated tumours which are larger than normal; *tmr* which causes tumours with a large number of roots; and *tms* which causes tumours having a large number of shoots (see Fig. 7.9). The effect appears to be brought about by the action of T-DNA gene products on the cytokinin/auxin

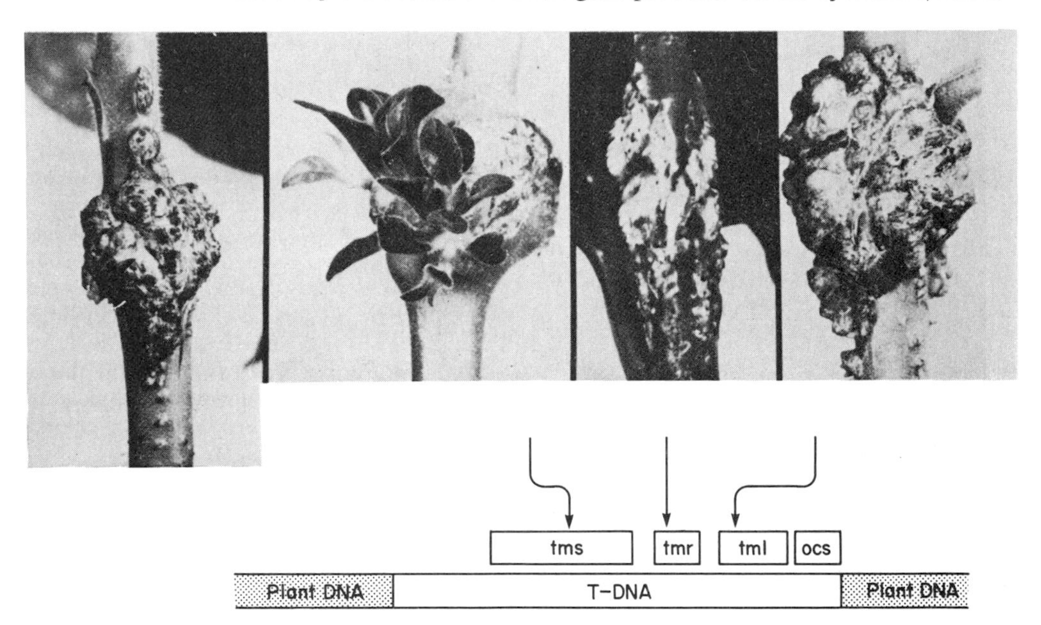

Figure 7.9 Loci within the T-RNA that control tumour morphology.

ratios within the plant tissue [42]. In this respect the correlation follows the pattern which can be observed when these ratios are changed for normal tissue in culture, i.e. shoot formation is favoured by high cytokinin/auxin ratios and root formation by low ratios. The function of *tml* is not clear, since these mutants appear to alter neither the ratios of the phytohormones nor their levels. Seven polyadenylated transcripts encoded by the octopine T-DNA have been identified, of which four have quite strong homology to the T-DNA of nopaline plasmids [43]. One transcript has been selected by hybridization to a restriction fragment and following translation been shown to encode octopine synthase. This transcript together with one other is specific to this type of Ti plasmid. Of the transcripts common to octopine and nopaline types of T-DNA, two clearly act to suppress shoot formation and one to suppress root formation.

7.3.3 T-DNA as a vector

The Ti plasmid has enormous potential as a vector for introducing foreign genes into plants. We have already seen that bacterial transposons inserted into T-DNA can be propagated in plant chromosomes and subsequently recovered having lost none of their functional capabilities. Furthermore, it has recently been shown that it is possible to regenerate plants from Ti plasmid induced tumours, in which the T-DNA is inherited in a Mendelian manner from one generation to another. This necessitates using a mutant Ti plasmid in which the tumourous properties are suppressed. One such mutant obtained by Tn7 mutagenesis produced tumours which gave rise to shoots. It proved possible to regenerate a plant from one of these shoots which had octopine synthase activity [44]. The progeny of this plant inherited the T-DNA and hence octopine synthase activity in all of their tissues. Similar experiments have been carried out with a nopaline plasmid in which the yeast alcohol dehydrogenase gene was inserted into the 'rooty' locus of the T-DNA [45]. Cells from the tumourous growth could be regenerated into complete tobacco plants which contained some 20 copies of the yeast gene in each of their cells. After self-pollination, the plants produced seeds which grew into plants still containing multiple copies of the yeast gene.

The yeast gene used in thc experiments described above was not expressed in thc tobacco plants. In order to achieve expression of foreign genes introduced into plants in this way, several groups have now placed foreign genes under the control of T-DNA promoters. Herrera-Estrella and coworkers [46], for example, have made use of knowledge which accrued from the detailed investigation of the promoter region of the nopaline synthase gene. They have made a recombinant Ti plasmid in which the genes for either octopine synthase or the bacterial enzyme chloramphenicol transacetylase (see also Chapter 8) are downstream of the nopaline synthase promoter. Tobacco plant cells transformed with these plasmids produce functional octopine synthase or chloramphenicol transacetylase. The group have gone on to use nopaline synthase promoter to direct the expression of two bacterial genes: the aminoglycoside phosphotransferase gene of Tn5 which inactivates the antibiotics kanamycin, neomycin and G418 (see also Chapter 8) and the methotrexate-insensitive dihydrofolate reductase of the R67 plasmid [47]. Since kanamycin, G418 and methotrexate are very toxic to plant cells, this forms the basis of a dominant selection system for plant cell transformation. These experiments clearly represent only the beginnings of the development of expression vectors based on the Ti plasmid. The future question will be of how to choose and isolate genes likely to improve crop plants to be introduced by this route. The obvious objectives of genes which give pest resistance or cold and drought tolerance will be difficult to attain since such genes will be difficult to recognize and isolate by recombinant DNA techniques. There are a number of more readily attainable objectives which could include the manipulation of genes for the sweet tasting plant protein Thaumitin, or genes for plant storage proteins engineered so that they have a better nutritional blend of amino-acids. One step in this direction has been taken by the introduction of the gene for the bean plant storage protein, phaseolin, into the cells of the sunflower plant [48]. The gene was transferred in a Ti vector and is expressed under the control of the octopine synthase promoter. One limitation of the Ti plasmid is its specificity for dicotyledonous plants, whereas most of the important food crops are the monocotyledonous grasses and cereals. Little is known of the factors which control this host-range specificity. Much basic research needs to be done on this and also on the mechanisms whereby plasmid DNA is transferred from the bacterium into the plant cell chromosome. Nevertheless Ti plasmid offers remarkable opportunities for crop improvement by genetic engineering techniques.

References

1. Hinnen, A., Hicks, J. B. and Fink, G. R. (1978) Transformation of yeast. *Proc. Natn Acad. Sci. USA*, **75**, 1929–33.
2. Struhl, K., Stinchcomb, D. T., Scherer, S. and Davis, R. W. (1979) High frequency transformation of yeast: autonomous replication of hybrid DNA molecules. *Proc. Natn Acad. Sci. USA*, **76**, 1035–9.
3. Scherer, S. and Davis, R. W. (1979) Replacement of chromosome segments with altered DNA sequences constructed *in vitro*. *Proc. Natn Acad. Sci. USA*, **76**, 4951–5.
4. Shortle, D., Haber, J. E. and Botstein, D. (1982) Lethal disruption of the yeast actin gene by integrative DNA transformation. *Science*, **217**, 371–3.
5. Neff, N. E., Thomas, J. H., Grisafi, P. and Botstein, D. (1983) Isolation of the β tubulin gene from yeast and demonstration of its essential function *in vivo*. *Cell*, **33**, 211–9.
6. Orr-Weaver, T. L., Szostak, J. W. and Rothstein, R. J. (1981) Yeast transformation: a model system for the study of recombination. *Proc. Natn Acad. Sci. USA*, **78**, 6354–8.
7. Rothstein, R. J. (1983) in *Methods in Enzymology* (eds. R. Wu, L. Grossman and K. Moldave). Academic Press, New York, London, 101–2.
8. Hsiao, C.-L. and Carbon, J. (1979) High frequency transformation of yeast by plasmids containing the cloned yeast *arg*4 gene. *Proc. Natn Acad. Sci USA*, **76**, 3829–33.
9. Stinchcomb, D. T., Struhl, K. and Davis, R. W. (1979) Isolation and characterisation of a yeast chromosomal replicator. *Nature*, **282**, 39–43.
10. Stinchcomb, D. T., Thomas, M., Kelly, J., Selker, E. and Davis, R. W.

(1980) Eukaryotic DNA segments capable of autonomous replication in yeast. *Proc. Natn Acad. Sci. USA*, **77**, 4559–63.
11. Clarke, L. and Carbon, J. (1980) Isolation of a yeast centromere and construction of functional small circular chromosomes. *Nature*, **287**, 504–9.
12. Murray, A. W. and Szostak, J. W. (1983) Construction of artificial chromosomes in yeast. *Nature*, **305**, 189–93.
13. Beggs, J. D. (1978) Transformation of yeast by a replicating hybrid plasmid. *Nature*, **275**, 104–9.
14. Beach, D. and Nurse, P. (1981) High frequency transformation of the fission yeast *S. pombe*. *Nature*, **290**, 140–2.
15. Beach, D., Durkacz, B. and Nurse, P. (1982) Functionally homologous cell cycle control genes in budding and fission yeast. *Nature*, **300**, 706–9.
16. Roggenkamp, R., Kusterman-Kuhn, B. and Hollenberg, C. P. (1981) Expression and processing of bacterial β galactomase in the yeast *S. cerevisiae*. *Proc. Natn Acad. Sci USA*, **78**, 4466–70.
17. Cohen, J. D., Eccleshall, T. R., Needleman, R. B. *et al.* (1979) The functional expression in yeast of the *Escherichia coli* plasmid gene coding for chloramphenicol acetyl transferase. *Proc. Natn Acad. Sci. USA*, **77**, 1078–82.
18. Jimenez, A. and Davies, J. (1980) Expression of a transposable antibiotic resistance element in *Saccharomyces*. *Nature*, **287**, 869–771.
19. Langford, C. J. and Gallwitz, D. (1983) Evidence for an intron contained sequence required for the splicing of yeast RNA polymerase II transcripts. *Cell*, **33**, 519–27.
20. McNeil, J. B. and Friersen, J. D. (1981) Expression of the herpes simplex virus thymidine kinase gene in *Saccharomyces cerevisiae*. *Mol. Gen. Genet.*, **184**, 386–93.
21. Hitzeman, R. A., Hagie, F. E., Levine, H. L. *et al.* (1981) Expression of a human gene for interferon in yeast. *Nature*, **293**, 717–22.
22. Valenzuela, P., Medina, A., Rutter, W. J., Ammerer, G. and Hall, B. D. (1982) Synthesis and assembly of hepatitis B virus surface antigen particles in yeast. *Nature*, **298**, 347–50.
23. Miyanohara, A., Toh-E, A., Nozaki, C. *et al.* (1983) Expression of hepatitis B surface antigen gene in yeast. *Proc. Natn Acad. Sci. USA*, **80**, 1–5.
24. Case, M. E., Schweizer, M., Kushner, S. R. and Giles, N. H. (1979) Efficient transformation of *Neurospora crassa* by utilising hybrid plasmid DNA. *Proc. Natn Acad. Sci. USA*, **76**, 5259–63.
25. Hughes, K., Case, M. E., Geerer, R., Vapnek, D. and Giles, N. H. (1983) Chimeric plasmid that replicates autonomously both in *Escherichia coli* and *Neurospora crassa*. *Proc. Natn Acad. Sci. USA*, **80**, 1053–7.
26. Stohl, L. L. and Lambovitz, A. M. (1983) Construction of a shuttle vector for the filamentous fungus *Neurospora crassa*. *Proc. Natn Acad. Sci. USA*, **80**, 1058–62.
27. Tilburn, J., Scazzocchio, C., Taylor, G. G. *et al.* (1983) (Personal communication).
28. Nester, E. W. and Kosuge, T. (1981) Plasmids specifying plant hyperplasias. *Ann. Rev. Microbiol.*, **35**, 531–65.
29. Van Larebake, N., Engler, G., Holsters, M. *et al.* (1974) Large plasmid in

Agrobacterium tumefaciens essential for Crown Gall inducing ability. *Nature*, **252**, 169–70.
30. Garfinkel, D. J. and Nester, E. W. (1980) *Agrobacterium tumefaciens* mutants affected in Crown Gall tumorigenesis and octopine metabolism. *J. Bact.*, **144**, 732–43.
31. Montoya, A. L., Chilton, M.-D., Gordon, M., Sciaky, D. and Nester, E. W. (1977) Octopine and nopaline metabolism in *Agrobacterium tumefaciens* and Crown Gall tumor cells: role of plasmid genes. *J. Bact.*, **129**, 101–7.
32. Chilton, M.-D., Saiki, R. K., Yadav, N., Gordon, M. P. and Quetier, F. (1980) T-DNA from *Agrobacterium* Ti plasmid is in the nuclear DNA fraction of Crown Gall tumour cells. *Proc. Natn Acad. Sci. USA*, **77**, 4060–4.
33. Willmitzer, L., De Beuckeleer, M., Lemmers, M., Van Montagu, M. and Schell, J. (1980) DNA from Ti plasmid present in nucleus and absent from plastids of Crown Gall plant cells. *Nature*, **287**, 359–61.
34. Thomashow, M. F., Nutter, R. L., Montoya, A. L., Gordon, M. P. and Nester, E. W. (1980) Integration and organisation of Ti plasmid sequences in Crown Gall tumours. *Cell*, **19**, 729–39.
35. Lemmers, M., De Benckeleer, M., Holsters, M. *et al.* (1980) Internal organisation boundaries and integrating of Ti plasmid DNA in nopaline Crown Gall tumours. *J. Mol. Biol.*, **144**, 353–76.
36. Holsters, M., Villarroel, R., Van Montagu, M. and Schell, J. (1982) The use of selectable markers for the isolation of plant-DNA/T-DNA junction fragments in a cosmid vector. *Mol. Gen. Genet.*, **185**, 283–9.
37. Yadav, N. S., Van Der Leyden, J., Bennett, D. R., Barnes, W. M. and Chilton, M. D. (1982) Short direct repeats flank the tDNA on a nopaline Ti plasmid. *Proc. Natn Acad. Sci. USA*, **79**, 6322–6.
38. Holsters, M., Villarroel, R., Gielen, J. *et al.* (1983) An analysis of the boundaries of the octopine Ti-DNA in tumours induced in *Agrobacterium tumefaciens*. *Mol. Gen. Genet.*, **119**, 35–41.
39. Simpson, R. B., O'Hara, P. J., Kwok, W. *et al.* (1982) DNA from the A6S/2 Crown Gall tumour contains scrambled Ti plasmid sequences near its junctions with plant DNA. *Cell*, **29**, 1005–14.
40. Garfinkel, D. J., Simpson, R. B., Ream, L. W. *et al.* (1981) Genetic analysis of Crown Gall: fine structure map of the T-DNA by site directed mutagenesis. *Cell*, **27**, 143–53.
41. Leemans, J., Deblaere, R., Willmitzer, L. *et al.* (1982) Genetic identification of functions of Ti-DNA transcripts in octopine Crown Galls. *EMBO J.*, **1**, 147–52.
42. Akiyoshi, D. E., Morris, R. O., Hinz, R. *et al.* (1983) Cytokinin/auxin balance in Crown Gall tumours is regulated by specific loci in the T-DNA. *Proc. Natn Acad. Sci. USA*, **80**, 407–11.
43. Willmitzer, L., Simons, G. and Schell, J. (1982) The Ti-DNA in octopine Crown Gall tumours codes for seven well defined polyadenylated transcripts. *EMBO J.*, **1**, 139–46.
44. De Greve, H., Leemans, J., Hernalsteens, J. P. *et al.* (1982) Regeneration of normal and fertile plants that express octopine synthetase from tobacco Crown Galls after deletion of tumour controlling functions. *Nature*, **300**, 752–5.

45. Barton, K. A., Binns, A. N., Matzke, A. J. M. and Chilton, M. D. (1983) Regeneration of intact tobacco plants containing full length copies of genetically engineered T-DNA and transmission of T-DNA to R1 progeny. *Cell*, **32**, 1033–43.
46. Herrera-Estrella, L., Depicker, A., Van Montagu, M. and Schell, J. (1983) Expression of chimaeric genes transferred into plant cells using a Ti plasmid derived vector. *Nature*, **303**, 209–13.
47. Herrera-Estrella, L., De Block, M., Messens, E. *et al.* (1983) (Personal communication).
48. Murai, N., Sutton, D. W., Murray, M. G. *et al.* (1983) Phaseolin gene from bean is expressed after transfer to sunflower via tumour inducing plasmid vectors. *Science*, **222**, 476–81.

8 Expression of cloned genes in animal cells

8.1 Direct transformation of mammalian cells

The ability to transfer DNA into mammalian cells and so complement a mutation has been known for over two decades [1]. The technique remained inefficient until Graham and van der Eb [2] developed an assay for the infectivity of viral DNA in which the DNA is applied to the cells as a coprecipitate with calcium phosphate. Several groups subsequently demonstrated that using this technique it was possible to transfer the thymidine kinase (*tk*) gene from herpes simplex virus (HSV) DNA into cells deficient for this enzyme. Such tk^- cells survive because the normal pathway for the synthesis of dTTP is from dCDP and not from the dTMP produced by the phosphorylation of thymidine by thymidine kinase. Tk^- cells do not survive in 'HAT' medium which contains hypoxanthine and aminopterin to block the synthesis of dTTP from dCDP and thymidine. Tk^- cells that acquire the HSV *tk* gene can therefore be selected by growth in HAT medium. The observation that carrier DNA greatly enhanced the transformation frequency led to successful attempts to transfer single copy eukaryotic genes using total genomic DNA [3]. Wigler and colleagues then demonstrated that unlinked, non-selectable genes could be cotransformed when mixed in excess with selectable DNA [4, 5]. In these experiments the herpes simplex virus *tk* gene was used as a selectable marker in the presence of a thousand-fold excess of ϕX174, pBR322 or rabbit β-globin gene sequences. If on the other hand the cells were plated on non-selective media following treatment with DNA, then it was not possible to detect any of these DNAs in random clones of cells. It seems therefore that selection identifies a population of cells capable of taking up considerable quantities of exogenous

DNA. Analysis of the biochemically transformed cells showed that the selectable marker gene becomes associated with high molecular weight DNA which could be the salmon DNA used as carrier [5] (Fig. 8.1). Alternatively when cells are transformed without carrier DNA then cell lines have been observed in which the transforming plasmid genome is present in multimeric form [6].

A number of techniques have been used to recover the selectable marker gene from the biochemically transformed cell lines. Perucho *et al.* [7] ligated pBR322 to chicken DNA and used the DNA to transform tk^- mouse cells. They isolated DNA from the resulting tk^+ clones, cleaved it with *Eco*RI and *Bam*HI, cyclized the fragments and used it to transform *E. coli* selecting for an antibiotic resistance marker carried by pBR322. In this way the chicken *tk* gene was isolated. A similar procedure has also been applied in order to clone the hamster gene for adenine phosphoribosyl transferase (APRT). The hamster DNA was cut with an enzyme that does not cleave within APRT and then ligated to pBR322. This DNA was used to transform $APRT^-$ mouse cells. DNA was later prepared from the

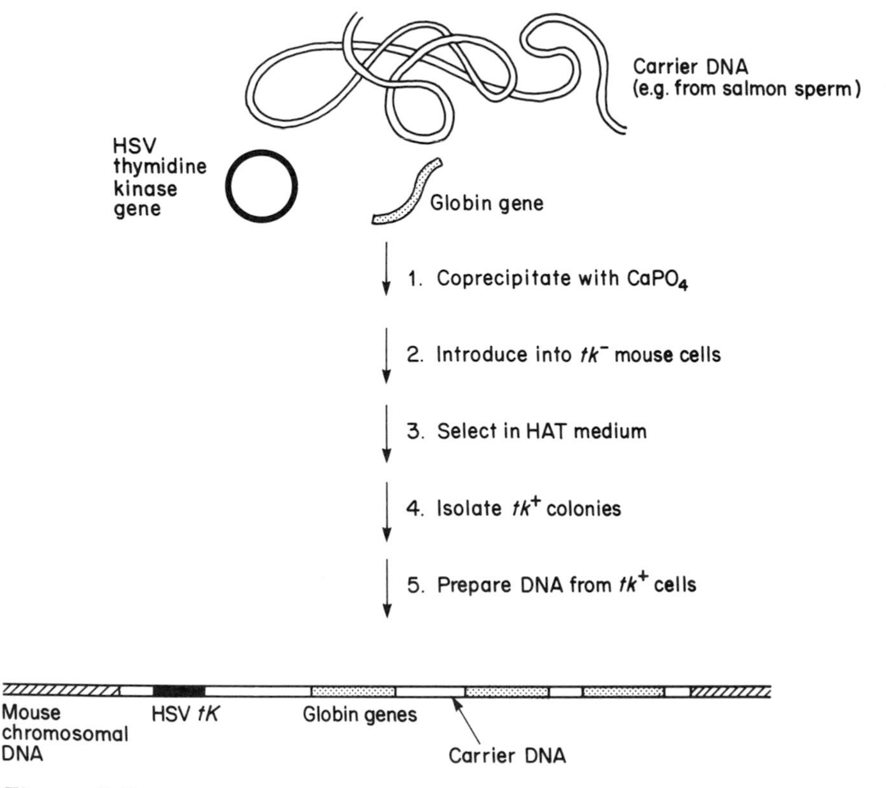

Figure 8.1 Cotransformation of tk^- mouse cells with the HSV *tk* gene, the rabbit β-globin gene and carrier DNA.

transformed *APRT*$^+$ mouse cells and made into a library in a bacteriophage λ vector. The library was then screened by hybridization with ^{32}P labelled pBR322 DNA. This identifies a clone containing the hamster *APRT* gene linked to pBR322 [8] (Fig. 8.2).

One of the most impressive applications of this approach has been

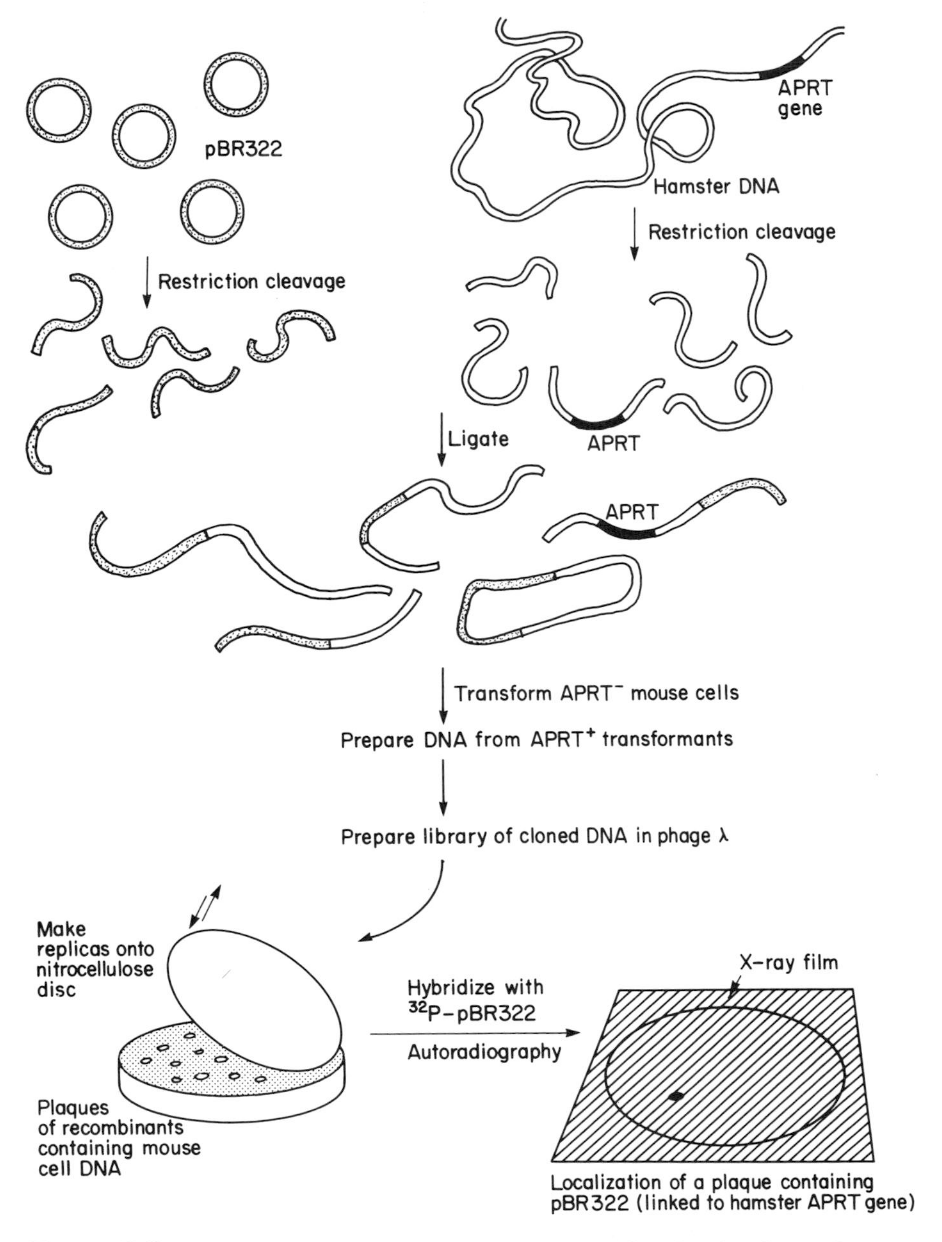

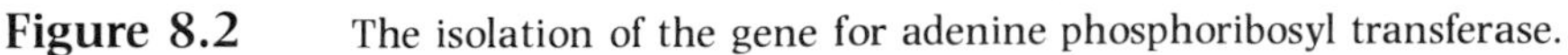
Figure 8.2 The isolation of the gene for adenine phosphoribosyl transferase.

the identification and eventual isolation of 'transforming genes' from human tumours *per se* and from cell lines derived from tumours. These DNA segments have the ability to alter or 'transform' the growth properties of mouse cells so that they form foci of cells which outgrow their non-transformed counterparts. (It is important to distinguish the specific use of the word 'transformation' in the sense of oncogenic transformation, from the general meaning of an alteration in any aspect of the genotype of a cell as a result of introducing new DNA.) Most of these experiments have been carried out by incubating DNA from human tumours with mouse cells and selecting foci of oncogenically transformed cells. Human DNA contains more than 10^5 copies of a repetitive element known as the '*Alu*' sequence because it contains the cleavage site of the restriction endonuclease *Alu*I. These sequences are dispersed throughout the genome and so there is an extremely high probability that any human gene is closely linked to such a sequence. A clonal cell line is established from a focus of transformed mouse cells and DNA extracted from this cell line is cloned into a bacteriophage library. The library is screened for human *Alu* sequences with a radiolabelled probe. The recombinant bacteriophage λ clones identified in this way contain the transforming gene from the human tumour linked to an *Alu* sequence (Fig. 8.3). An alternative approach has been to join fragments of DNA from a human tumour cell line to a prokaryotic marker gene. This DNA is then used to transform mouse cells. The transformed foci are picked and cultured in order that sufficient chromosomal DNA can be prepared for cloning in bacteriophage λ. A prokaryotic marker that has been used is the *E. coli* tRNA suppressor gene, *sup*F. In this case the phage vector contains amber mutations and will not grow in a suppressor-free bacterial host (see Chapter 4). Thus the only phage that will grow on such a host are those in which both amber mutations have reverted, or those which have acquired the *sup*F gene linked to the transforming gene from the transformed mouse cells (Fig. 8.4). These techniques have enabled transforming genes to be isolated from bladder cancers, neuroblastomas and colon cancers [9, 10]. These genes are similar to the transforming genes found in retroviruses (see Section 8.5). The human bladder cancer gene, for example, corresponds to the *ras* oncogene of Harvey sarcoma virus.

8.2 SV40 vectors

Although DNA can be introduced into cells by the techniques discussed in the previous section, there is a clear need to utilize viral replicons together with selective systems in much the same way that we have seen for bacterial cloning systems. The detailed knowledge of

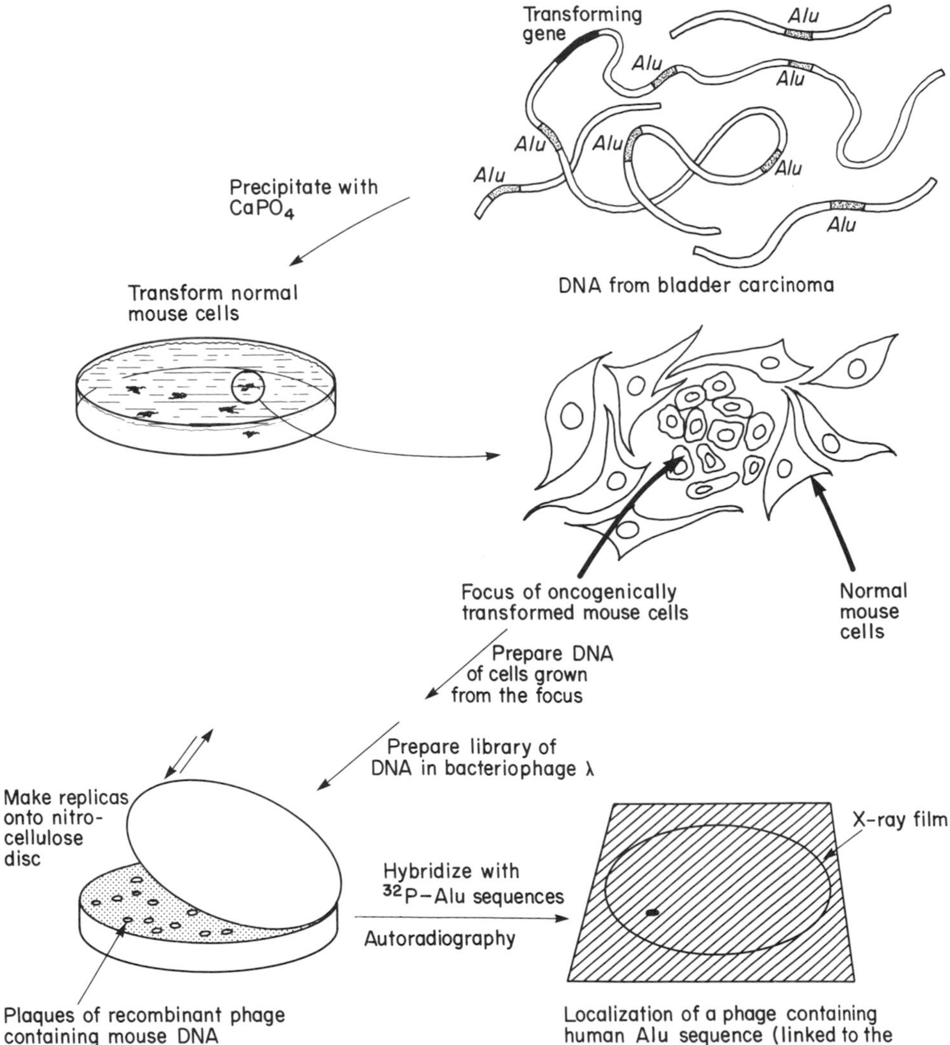

Figure 8.3 Cloning an oncogene linked to *Alu* sequences.

the molecular biology of the small DNA tumour virus, simian virus 40 (SV40) [11], has made it a primary candidate for development as a vector for introducing foreign DNA sequences into cultured mammalian cells. The virus can have two kinds of interaction with its host cell, depending upon the species. In permissive monkey cells one sees a productive infection cycle resulting in cell death and the release of progeny virions. In non-permissive mouse or rat cells a small proportion of cells become transformed undergoing heritable alterations in their growth properties. Such transformed cells will produce

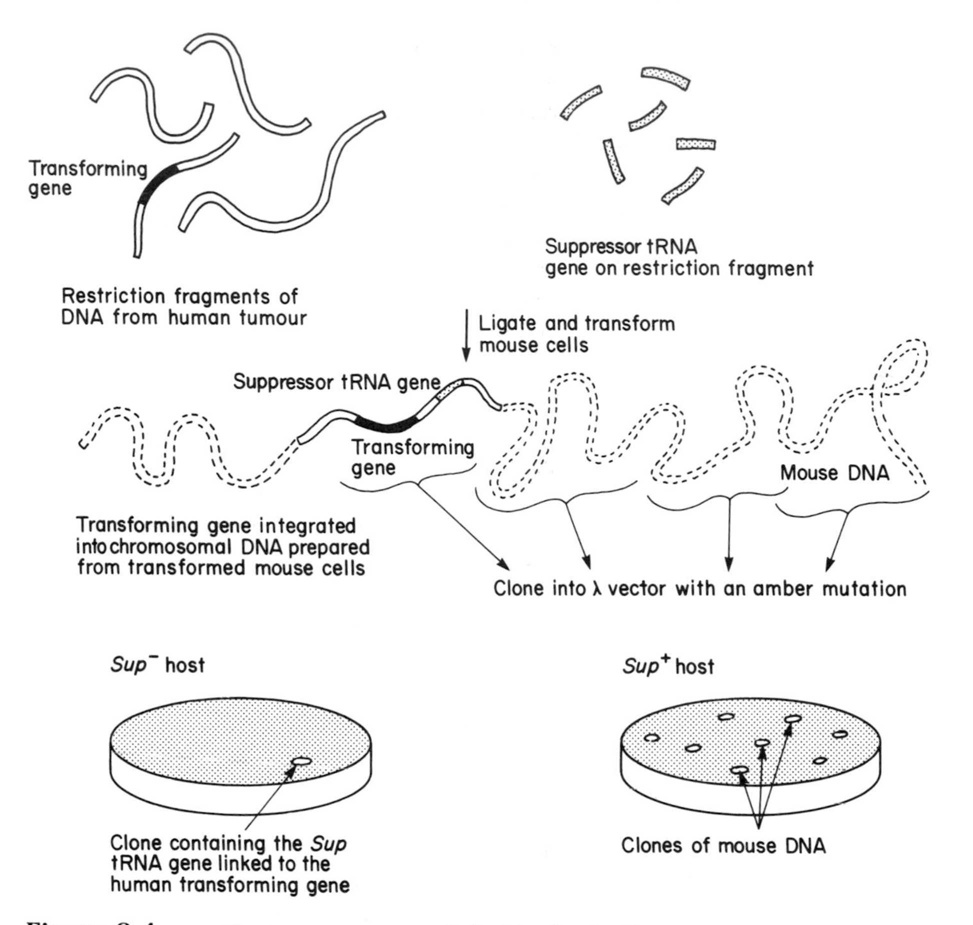

Figure 8.4 Cloning an oncogene linked to the *SupF* gene.

tumours when injected into immuno-incompetent animals. In non-permissive infections only the early viral genes are efficiently expressed, and in those cells which become stably transformed the viral DNA becomes integrated into the host cell chromosomes. In the infection of permissive cells the virus is first transported to the nucleus where it is uncoated. Stable early transcripts are produced which correspond to about half of the genome (see Fig.8.5). Complementation analysis of temperature-sensitive mutants indicates a single early complementation group, A. The product of this gene, the viral large tumour antigen (T-ag), is required to initiate DNA replication which takes place bidirectionally from the single origin shown in Fig. 8.5. The early messenger RNAs code for a second tumour antigen, a smaller related protein called the small tumour antigen (t-ag). The two early genes contain non-coding intervening

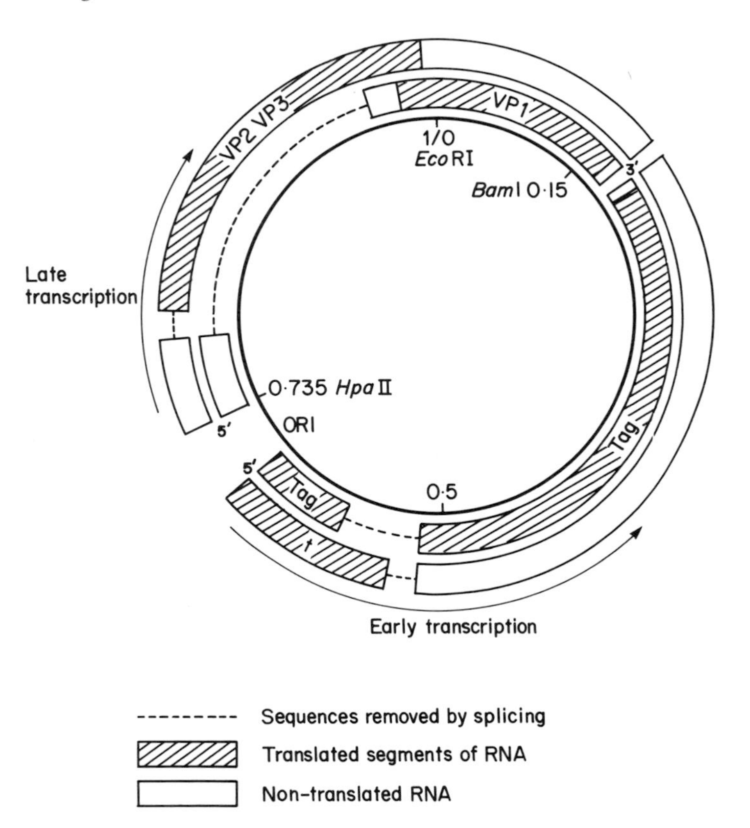

Figure 8.5 Early and late transcriptional circuits of SV40.

sequences which are removed from the mRNA by splicing. A larger intervening sequence is removed from the T-ag RNA which therefore becomes the smaller message. Only some of these intervening sequences are removed from the small t-ag mRNA so this RNA molecule is larger. It encodes a smaller polypeptide, however, since it contains a termination codon in that part of its sequence not found in T-ag mRNA. Late transcripts are found after the onset of viral DNA replication and these encode the three capsid proteins VP1, VP2 and VP3. These mRNAs are transcripts of the opposite strand to the early mRNA and they each have 5′ leader sequences which correspond to sequences on the genome which are well-removed from the coding sequence. The intervening sequences between the leaders and the body of the mRNA are again removed by splicing.

8.2.1 Defective viral vectors

As with the case of bacteriophage λ, there is a physical limit to the amount of DNA which can be packaged into the virus capsid. Since the SV40 genome is small (5.2 kb) and since the three identified non-essential regions of the genome only amount to several hundred base pairs of DNA the use of non-defective viral vectors is limited. This has led to the use of defective viral vectors whose infection process has to be 'helped' by viral DNA which supplies the missing functions. The helper viruses carry a temperature-sensitive mutation complemented by the segment of DNA within the vector [12]. The scheme for the use of such vectors is shown in Fig. 8.6 for the specific case of SVGT-1, the *Bam*HI/*Hpa*II fragment of SV40 DNA which carries the

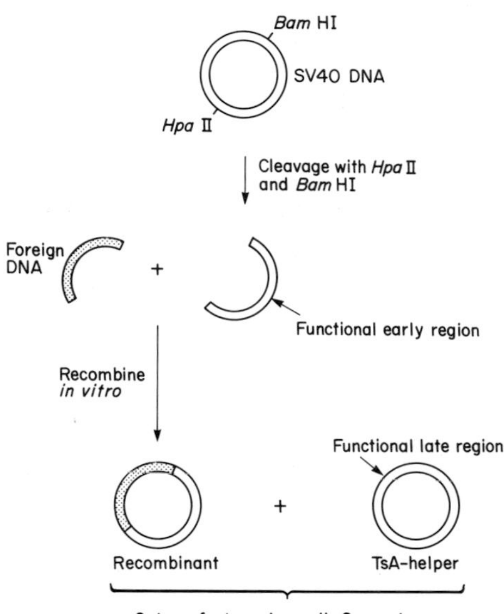

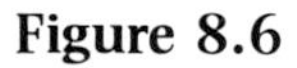

Figure 8.6 The use of a helper virus to complement a defective SV40 vector carrying foreign DNA.

early genes and replication origin. Recombinants between this DNA sequence and foreign DNA are introduced into monkey cells together with DNA from the *tsA* mutant helper virus which has a temperature sensitive mutation in the early complementation group. The infection of cells with the two virus types is then detected by the formation of plaques at 41° C, the non-permissive temperature for the helper. The recombinant virus therefore provides the early functions and the helper virus the late functions for the infective cycle.

8.2.2 Splicing and mRNA stability

The analysis of the transcription of a number of genes cloned into vectors such as SVGT-1 [12] indicated that, although the foreign segments of DNA were transcribed, discrete functional RNA molecules were not seen. Mulligan *et al.* [13] argued that this was due to the deletion in these vectors of the viral DNA sequences required for processing late mRNA; the late leader sequence and the sequence in the body of the VP1 gene to which this is spliced. In order to leave these sites intact, they developed a vector, SVGT-5, which lacks sequences between the *Hin*dIII site and *Bam*HI site found internally within the sequence of the body of the VP1 gene. They used this vector to clone a segment of rabbit β-globin cDNA containing the Met-initiating codon 37 nucleotides from its *Hin*dIII terminus and the translation termination codon just proximal to its *Bgl*II terminus. When the recombinant was introduced into monkey cells together with helper SV40 *tsA* DNA, it was possible to detect discrete 1.8 and 1.0 kb RNA molecules which contained both SV40 leader sequences and β-globin coding sequences. Furthermore, these cells were demonstrated to be synthesizing a protein immunologically recognizable as rabbit β-globin.

Hamer and Leder cloned the chromosomal gene for the mouse β-globin gene between the *Bam*HI and *Hpa*II sites of SV40 DNA. The chromosomal segment contained the intact β-globin gene with both its intervening sequences, together with an additional 430 residues to the 3′ end. This recombinant virus directs the production of stable globin mRNA which has the SV40 leader sequences and is approximately 300 bases longer than authentic globin mRNA. The intervening sequences are apparently correctly removed from the mouse gene by the monkey processing enzymes and furthermore, the cells synthesize a polypeptide recognizable as globin both immunologically and by tryptic fingerprint analysis. In these cases the production of unstable RNAs can be specifically correlated with the absence of splicing sequences. Recombinants have been constructed in which both the position of the globin gene and the number of available splice junctions have been varied. These studies suggest that splicing activity is necessary for the production of the stable RNA. Any splice junction will suffice whether it be an SV40 specified sequence or sequences within the cloned gene [14]. A similar conclusion was reached by Gruss *et al.* [15] who cloned a cDNA copy of VP1 mRNA into an SV40 replacement vector to give an 'intronless' late transcription unit. This recombinant does not direct the synthesis of VP1. VP1 synthesis is restored, however, when the intron from the mouse β-globin gene is inserted into the 'intronless' VP1 gene [16]. The generality of the requirement for splicing is not certain. Gruss *et al.* [17] have constructed SV40 recombinants containing the rat pre-proinsulin gene and find efficient synthesis of rat pre-proinsulin whether or not the intron of the pre-proinsulin gene is present. In their construct lacking an intervening sequence, the pre-proinsulin gene is transcribed from the late promoter but the transcripts do not have the 5′ terminus that is normally the most abundant. Instead the transcripts have unusual 5′ ends, identical to those present on unspliced RNA made by a viable SV40 late deletion mutant. It could be therefore that these novel 5′ ends provide a sequence in the RNA that substitutes for the splicing requirement.

The gene that encodes the haemagglutinin (HA) glycoprotein of influenza virus has also been expressed in SV40 vectors. The late gene replacement vector that was used in these studies contains none of the splice donor and receptor sites used in the synthesis of late SV40 mRNA and yet the HA gene seems to be efficiently translated from the unspliced message. These experiments are also an interesting study on the requirements for the export and modification of this glycoprotein. The influenza HA becomes anchored into the surface membrane of infected cells. It has hydrophobic amino-acids at its *N*-terminal and *C*-terminal regions. In cells that are expressing the HA gene cloned in SV40, the protein becomes glycosylated and it is transported to the membrane where it is biologically active and can agglutinate erythrocytes. When the DNA sequences that encode the *C*-terminal region are deleted the protein no longer becomes anchored into the cell membrane. It still undergoes glycosylation, however, but is transported across the membrane and secreted into the culture medium. When the *N*-terminal coding sequences are deleted, the protein is neither glycosylated nor transported and remains intracellular. This suggests that protein export and glycosylation by enzymes associated with the endoplasmic reticulum are a coupled process [18]. This ability of animal cells to correctly modify proteins expressed from recombinant DNA molecules is a major advantage and is likely to lead to the increased use of these vectors in expression systems.

8.2.3 Mammalian cell *E. coli* shuttle vectors

E. coli still provides a highly versatile environment for the manipulation of cloned DNAs. The ease whereby DNA molecules can be reconstructed, cloned and propagated to give high DNA yields has meant that *E. coli* cloning systems are usually used to prepare recombinant molecules for use in mammalian cells. With this application in mind, a number of laboratories have developed shuttle vectors that are capable of replicating either in *E. coli* or in mammalian cells. One alternative SV40 cloning system uses the host monkey cell to supply the helper function that permits replication of DNA carrying the SV40 replicon (Fig. 8.7). This system has facilitated the development of plasmid vectors which can replicate

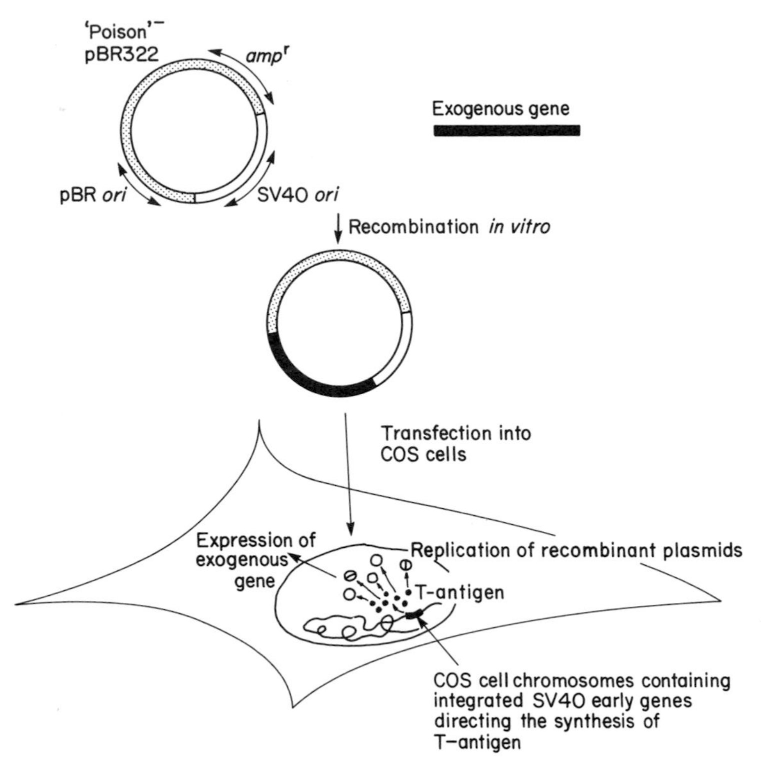

Figure 8.7 The use of COS cells to provide the helper function for defective SV40 vectors.

either in monkey cells or in *E. coli*. The host monkey cells are known as COS cells. They have been transformed by a plasmid carrying the early region of SV40 with a defective replication origin [19]. COS cells express wild type large T antigen and can support the growth of SV40 with deletion mutations in the early region. Recombinants containing foreign DNA inserted into the early region can therefore be propagated without the need of a helper virus. Since the only cis-acting function required for SV40 DNA replication is an 85 base pair DNA segment surrounding the replication origin, a series of plasmids have been constructed which contain this DNA segment and replicate in COS cells [20]. The efficiency of the replication depends upon the plasmid used. When pBR322 is linked to this 85 base pair segment and introduced into COS cells, part of the plasmid genome is deleted to form derivatives which replicate more efficiently [21]. It appears that pBR322 contains a sequence located in the deleted region which prevents it from replicating efficiently from the SV40 origin in monkey cells. This 'poison sequence' is not found in a number of derivatives of pBR322 including pBR327, pBR328 and pAT153. These all replicate efficiently in COS cells from an SV40 origin. Foreign DNA introduced into such molecules is replicated from the SV40 origin to greater than 10^5 copies per cell. The human α-globin gene introduced into COS cells in this way is efficiently expressed although expression of the human β-globin gene is much less efficient. The reasons for these differential levels of expression are not yet clear [22].

Berg's laboratory have developed a series of SV40 vectors which contain sequences derived from pBR322 and which can be propagated in *E. coli* [23]. The first of these, pSVGT5, consists of SVGT5 cloned into the *Pst*I site of a plasmid pSV1, which also contains a segment of the SV40 genome (Fig. 8.8). An SVGT5 recombinant containing a foreign gene can be cleaved from pSV1GT5 by *Pst*I digestion whereupon it may be introduced into monkey cells. Alternatively because the molecule contains a duplicated segment of the SV40 genome, it can be introduced directly into monkey cells and recombination across the duplication generates the equivalent of the SVGT5-recombinant.

Another series is exemplified by pSV2-X. This contains the

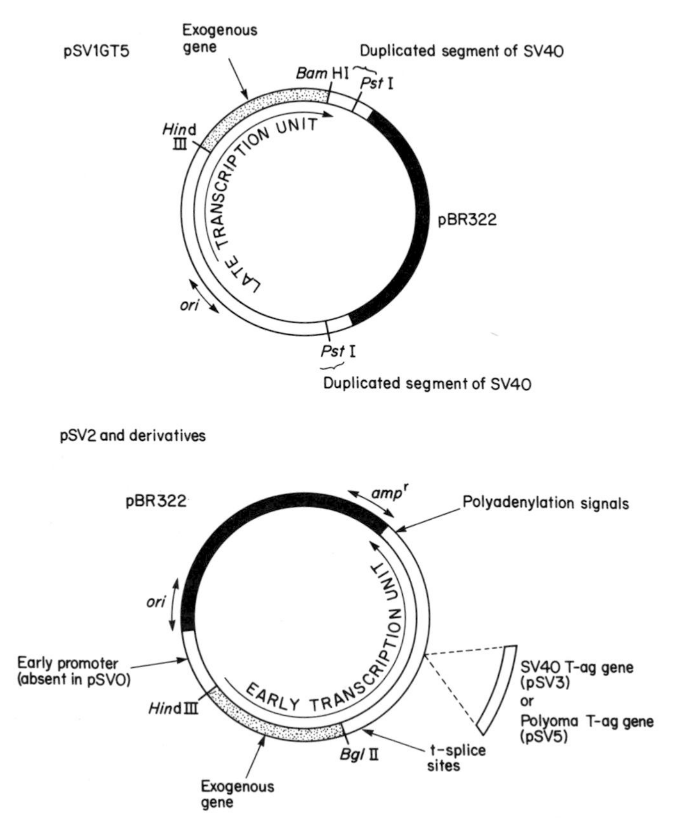

Figure 8.8 The pSV series of vectors.

bacterial replication origin and ampicillin-resistance gene of pBR322 linked to the replication origin and early promoter of SV40. Exogenous DNA can be inserted between a *Hin*dIII site and a *Bgl*II site downstream from the SV40 promoter. Further downstream from this insertion site is a segment of SV40 DNA containing the small t-antigen splicing site and another having the early polyadenylation signal. These should provide all the functions necessary for the maturation of messenger RNAs transcribed from the early viral promoter. There are a number of derivatives of pSV2-X. One of these, pSVO-X, lacks the segment containing the SV40 promoter. It therefore permits an assay for promoter function in the exogenous gene. Two other derivatives, pSV3-X and pSV5-X contain additional segments having the replication origins and entire T-antigen coding regions from SV40 or polyoma viral DNAs respectively (Fig. 8.8).

SV40 expression vectors containing promoter, splicing and polyadenylation sequences have proved useful for the expression of cDNA fragments. The cDNA cloning technique of Okayama and Berg (Section 2.2.2) provides an efficient means of cloning full-length cDNA. A number of vectors have therefore been modified so that they can be used in this method. Breathnach and Harris [24], for example, have described vectors in which the SV40 early or late promoters are linked to a pair of splice sites from the rabbit β-globin gene. Such plasmids are suitable for the Okayama and Berg cDNA cloning approach and are extremely powerful expression vectors.

8.2.4 SV40 vectors with selectable markers

The latest stage in the development of SV40 vectors has been to incorporate the gene for a selectable marker into vectors which only have the viral DNA replication functions. Such a selection system is of obvious advantage considering the inefficiency by which DNA can be introduced into mammalian cells.We have seen earlier in this chapter that the thymidine kinase gene of herpes simplex virus has found widespread use as such a selectable marker. It has the disadvantage, however, that it can only be used in mutant tk^- cells. Two dominant selection systems have been developed, both of which utilize prokaryotic genes.

The first of these revolves around the pathway for purine metabolism (Fig. 8.9), and utilizes the *E. coli gpt* gene which encodes xanthine–guanine phosphoribosyl transferase (XGPRT) which is equivalent to hypoxanthine–guanine phosphoribosyl transferase (HGPRT) in mammalian cells. When cloned into a number of SV40-based vectors and introduced into monkey cells, the gene is expressed. The gene product is distinguished from the endogenous mammalian enzyme both by its electrophoretic mobility and its lack

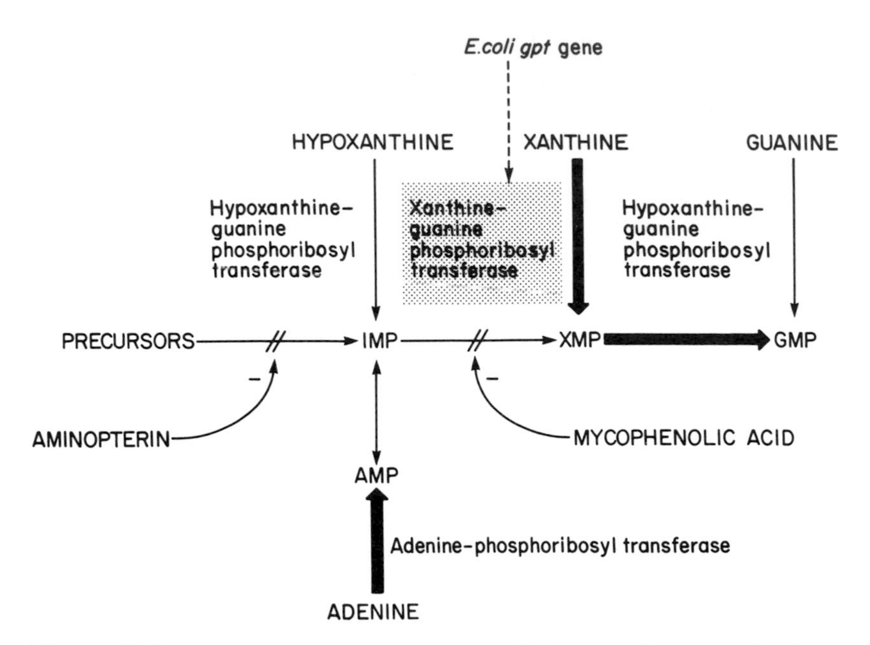

Figure 8.9 Pathways for purine metabolism to illustrate the direct selection for mammalian cells expressing the *E. coli gpt* gene in medium containing xanthine, adenine, mycophenolic acid and aminopterin.

of inhibition by hypoxanthine [25]. Secondly, there are lines of human cells which have been established from patients with Lesh–Nyhan syndrome, a defect in HGPRT. These cells are incapable of growth in medium containing hypoxanthine, aminopterin and thymidine (HAT medium). The cells can, however, be transformed with pSV-*gpt* DNA and those which express the *gpt* gene selected by their growth in HAT medium. The expression of *gpt* in mammalian cells has enabled a selection system to be developed for cloning [26]. This utilizes mycophenolic acid which blocks the conversion of IMP to XMP and so prevents the synthesis of GMP. This can be made more effective by the addition of aminopterin which prevents the synthesis of IMP (see Fig. 8.9). Mammalian cells will not grow in medium containing these two inhibitors supplemented with adenine and xanthine. The mammalian HGPRT cannot convert xanthine to XMP and hence GMP cannot be synthesized. If the *gpt* gene is introduced into the mammalian cell growth can occur in this medium and so such cells can be selected. The expression of *gpt* has been examined in cells infected with pSV1GT5-*gpt* and pSV1GT7-*gpt*, and an aberrant splicing pattern observed in which the acceptor site is located in the prokaryotic DNA. In cells transformed with pSV3-*gpt*, on the other hand, the *gpt* gene is transcribed from the SV40 promoter and the transcripts are processed as expected [27].

The second system utilizes the prokaryotic gene which encodes amino glycoside 3′ phosphotransferase. It was originally shown for yeast cells (see Section 7.1.3) and subsequently for mammalian cells [28] that this enzyme would endow resistance to the antibiotic G418. The latter experiments utilized a plasmid in which the aminoglycoside 3′ phosphotransferase gene was under the control of the HSV *tk* gene promoter. G418 resistant clones could be selected from cells transfected with such a plasmid, and these cloned cells shown to be expressing the bacterial enzyme. A plasmid carrying the G418 resistance gene could also be cotransfected into tk^- cells together with a second plasmid carrying the HSV *tk* gene. Of the cells selected by resistance to G418, 45% had become tk^+ and so had taken up both plasmids. The G418 resistance marker has also been successfully used in the pSV series of vectors to facilitate the introduction of a human β-interferon gene into mouse and rabbit cells [29].

We saw in Chapter 5 how the expression of the mammalian gene for dihydrofolate reductase (DHFR) had been engineered in *E. coli*. In this case a direct selection system could be employed since the mammalian and bacterial enzymes are differentially sensitive to trimethoprim (see Section 5.4). A selection system can also be employed with mammalian cells since some $dhfr^-$ cell lines are available, which absolutely require thymidine, glycine and purines for growth unless an exogenous gene is supplied. This was first shown for the cloned cDNA of the mouse *dhfr* gene inserted into SVG5 and SVGT7 vectors [30].

It is possible to increase the copy number of the *dhfr* gene in mammalian cells by gradually increasing the concentration of the anti-folate drug methotrexate over many cell generations. Wild type cells are killed by methotrexate. Resistance arises by amplification of the gene together with substantial flanking chromosomal regions. Kaufman and Sharp [31] have discovered that by linking a foreign gene to the *dhfr* gene and introducing this into cells, it is possible to amplify both genes by selection on media containing methotrexate. The plasmid which they used contained the *dhfr* gene linked to the splicing signals of an immunoglobulin gene and also an intact SV40 genome. The plasmid was introduced into $dhfr^-$ cells and cells selected in which DHFR was being expressed. These were then subjected to selection on media containing methotrexate and ultimately cloned lines were isolated in which gene amplification had occurred. In one such cloned line 10% of the soluble protein was of a polypeptide related to the SV40 small t-antigen.

8.3 Adenovirus vectors

Just as bacteriophage λ was known to be naturally capable of transducing genes of *E. coli*, it has for many years been known that the human adenovirus genome can carry exogenous genes. Specifically a number of natural recombinants between adenovirus DNA and SV40 DNA were isolated during the production of adenovirus vaccine in monkey cells contaminated with SV40. Our knowledge of the molecular biology of adenovirus is now such that the rational design of *in vitro* recombination experiments with adenovirus DNA is possible. These vectors could offer great potential as eukaryotic expression systems, since the virus lends itself quite well to large-scale propagation and furthermore during the course of its infection cycle the virus switches off the host cell protein synthesis thereby maximizing the synthesis of viral products.

Adenovirus serotypes 2 and 5 (Ad2 and Ad5) have been the most extensively studied. The virion contains a linear 35 kb DNA molecule with inverted terminal repeats. The 5′ ends of the DNA are covalently attached to a 55 k dalton protein which plays a critical role in DNA replication. In order to clone those terminal regions it is first necessary to remove the protein. The presence of the protein does, however, confer a higher degree of infectivity upon the DNA. Early in infection the viral DNA is expressed from five transcription units (Fig. 8.10). The majority of the late RNAs initiate at map position 16.6 and are processed from a complete transcript of the genome to the right of this position. The pattern of splicing and polyadenylation is highly complex, but each mature late RNA contains the 'tripartite leader' (the segments designated 1, 2 and 3 in Fig. 8.10) linked to one of the indicated polypeptide encoding regions.

As we have seen with bacteriophage λ, it is important to reduce the number of certain restriction sites before useful cloning vehicles can be constructed. This can be achieved by a procedure analogous to that used with bacteriophage λ. Viral DNAs are selected that are resistant to restriction endonucleases. Rather than imposing selection within the bacterial cell as could be done in selecting phage λ variants, the selection of adenoviral DNAs lacking restriction sites has been carried out *in vitro*. Variants of Ad5 DNA lacking *Xba*I sites, for example, were selected by cleaving the DNA with *Xba*I and then religating the fragments. Those fragments that are regenerated by ligation to intact viral genomes can be selected by virtue of their infectivity. If there are multiple sites for the restriction endonuclease, then the regeneration of the intact genome will occur at a low frequency. If, however, there are variants within the population which have lost a restriction site, then these will regenerate intact genomes at a higher frequency during the ligation step, and will subsequently be enriched [32].

The clustering of the E1 early genes near the left terminus

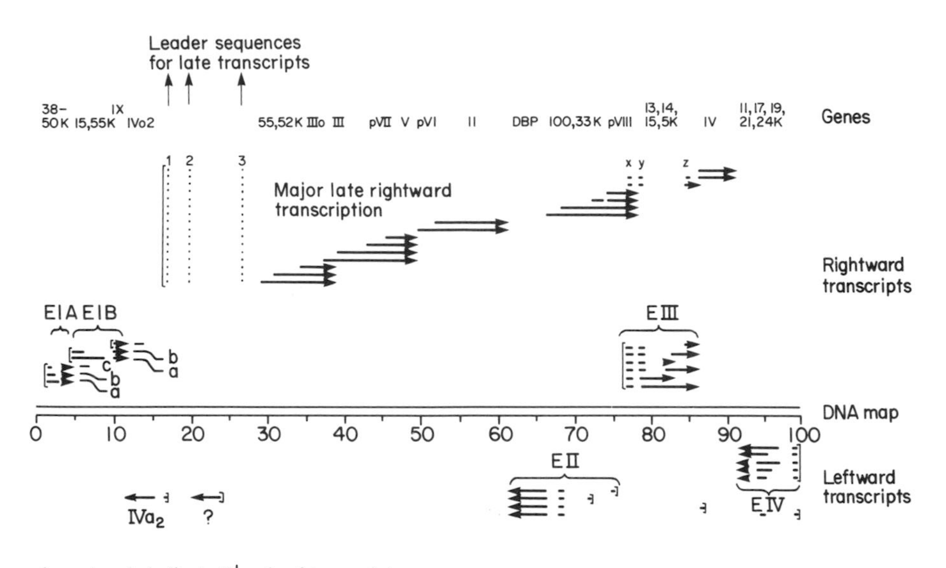

Figure 8.10 Adenovirus transcription units.

encourages the possibility of using this region as a cloning site, since it appears that adenoviruses have a mechanism for restoring the structure of one terminal repeat as long as the other remains intact. It has been shown, for example, that DNA from the left terminus of adenovirus 2, can be cloned and mutated *in vitro* and then ligated to a large restriction fragment comprising the rest of the Ad5 genome (Fig. 8.11). The segment of the Ad5 genome that has been used in such experiments was from the deletion mutant dl309 which has a single *Xba*I site. Upon transfection into human cells the left terminal region is correctly regenerated, and the progeny viral DNA is infective but contains the mutation introduced into the Ad2 segment during the cloning manipulations in *E. coli*. It turns out that it is not necessary to remove the plasmid vector sequences since these are removed by the cell and the adeno-terminal repeats restored in the infective cycle [33, 34].

There is considerable potential for expressing foreign DNA inserted into this region of adenovirus DNA. Several mutations have been introduced into the E1 transcription unit in this way to give viruses which cannot grow in human HeLa cells unless a helper function is provided. Similarly if foreign DNA were to be inserted into the E1 region by such a means, then a helper function would have to be provided in order to achieve the infective cycle. This can be provided by a human cell line (293 cells) which has been transformed by adenovirus DNA fragments and which expresses E1 functions [35].

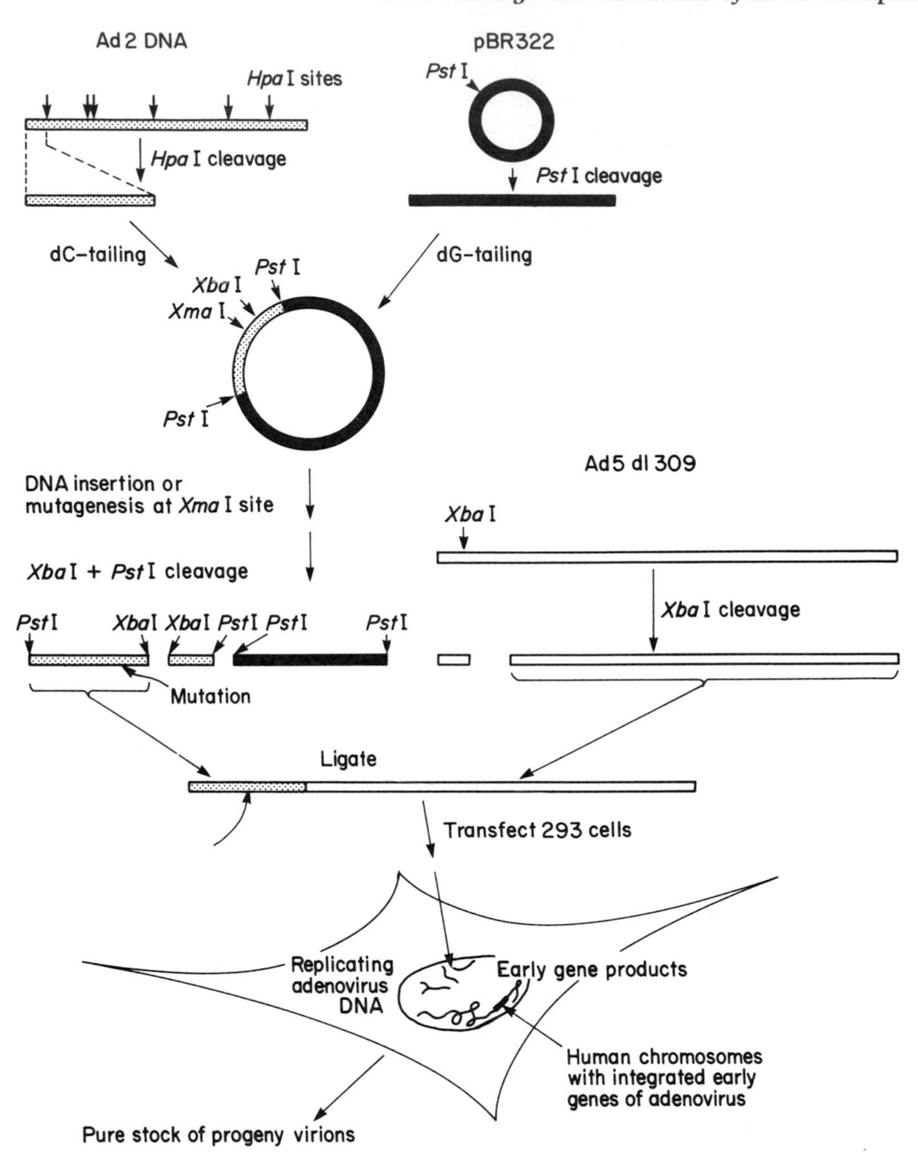

Figure 8.11 Procedure for manipulating the E1 region of the adenovirus genome in *E. coli*.

The use of adenovirus as a cloning vehicle is at an early stage. The recombinants that have so far been described contain the SV40 large T-antigen gene, the expression of which can be selected in monkey cells. Adenovirus infections of monkey cells are normally abortive but the block can be overcome by a helper effect of the SV40 large T-antigen. The 'natural' adeno-SV40 hybrids were in fact selected in this way. Recombinants have been constructed *in vitro* in which a restriction fragment of SV40 containing the entire large T-antigen coding region expressed under the control of the adenovirus major late promoter [36, 37]. We can expect to see the continuing development of such vectors.

8.4 Bovine papilloma virus vectors

The papilloma viruses are another group of viruses whose potential is only just being realized, especially as expression vectors. The papilloma viruses are responsible for causing warts and other tumours which are usually benign. Their molecular biology has lagged behind that of other viruses since there is no suitable tissue culture cell system for the propagation of infectious virions. It is, however, possible to use the DNA of bovine papilloma virus (BPV) to transform the growth properties of cultured cells. The viral genome is maintained episomally in the transformed cell [38, 39]. The transforming ability resides on a restriction fragment that comprises 69% of the BPV-1 genome [40]. This fragment has been used as a vector to introduce the rat pre-proinsulin gene into mouse cells [41]. The recombinant molecules are maintained episomally, each cell containing about 100 plasmid molecules. Proinsulin is efficiently expressed and secreted into the medium. Efficient expression of the human growth hormone gene carried on BPV-1 DNA has also been observed [42]. In this case the growth hormone gene was linked to the promoter of the mouse metallothionein gene and was inducible by heavy metals (see Section 8.7).

Attempts have also been made to construct BPV shuttle vectors capable of replication both in *E. coli* and in mouse cells. Unfortunately, however, a recombinant between pBR322 and either the whole of the BPV genome or a fragment containing the transforming gene has a reduced efficiency of focus formation in the transformation assay. This was not due to the 'poison' sequence that inhibits the replication of SV40–pBR322 recombinants (see Section 8.3) since deletion of this sequence did not relieve the block to BPV transformation [43]. In the course of these experiments, Di Maio and coworkers made the unexpected finding that by incorporating the human β-globin gene into this construct they could stimulate the transformation frequency by about 500-fold. The stimulatory effect is apparently due to a segment of human DNA that includes the third exon and some of the 3′ flanking sequences of the β-globin gene. The resulting plasmids can be maintained either in mouse cells or in *E. coli*. The human β-globin gene is expressed at high levels in mouse cells from its own promoter.

In order to get around the problem of the inhibitory effect of pBR322 sequences on transformation, these sequences are removed from the segment of the BPV genome linked to the foreign genes which is then introduced into mouse cells. In experiments of this type that were carried out to analyse the expression of the β-interferon gene, it was found that in different cell lines the induction of β-interferon expression with poly(I)–poly(C) varied by about ten-fold. This was thought to be due to rearrangement of sequences in the linear transforming DNA. Zimm *et al.* [44] have therefore cloned the β-interferon gene into a BPV vector containing the segment of the human globin gene that stimulates transformation and which has single receptor sites for *Bam*HI and *Sal*I fragments (Fig. 8.12). They

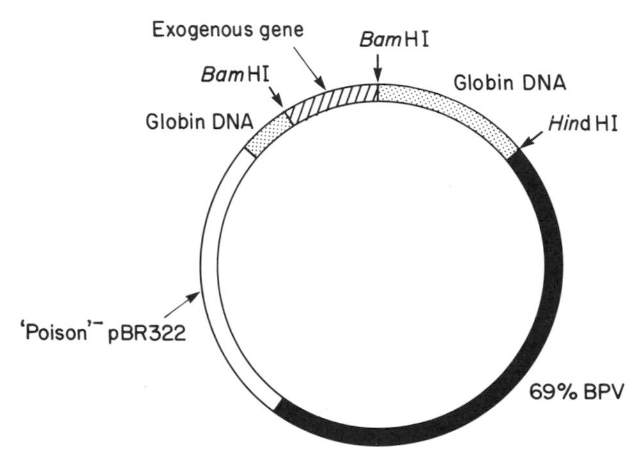

Figure 8.12 A bovine papilloma virus shuttle vector that contains a segment of the human β-globin gene.

observe consistent levels of expression in independent cell lines transformed by recombinants made with this vector. The β-interferon mRNA is induced by 400-fold when the cells are treated with poly(I)–poly(C). These recombinants have enabled two regulatory regions upstream of the gene to be identified. The deletion of one of these leads to a reduced level of both constitutive and inducible expression, whilst deletion of the other leads to elevated levels of constitutive expression.

8.5 Retroviral vectors

Clear and concise accounts of the molecular biology of retroviruses may be found in reference [45]. These viruses have single-stranded RNA genomes containing basically three genes: *gag*, which encodes viral core proteins; *pol*, which encodes reverse transcriptase; and *env*, which encodes the proteins of the viral envelope (Fig. 8.13). Such a virus will not transform cultured cells. Cellular transformation is

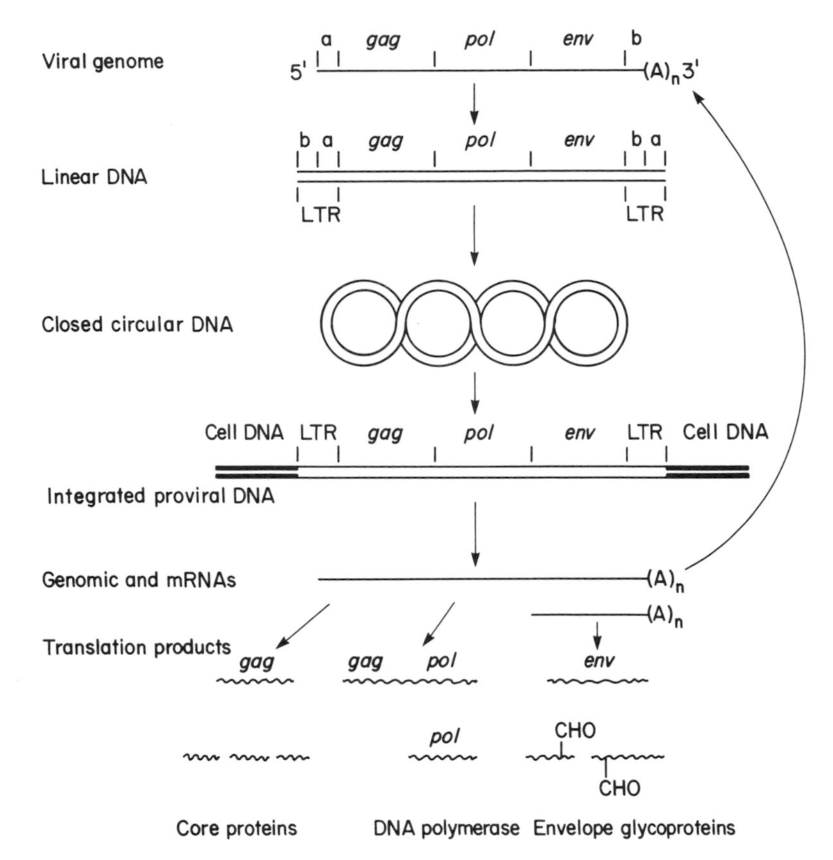

Figure 8.13 The replication cycle of retroviruses.

brought about by another gene designated *onc* for oncogene. In Rous sarcoma virus (RSV) for example, this additional gene is known as *src* and also causes the virus to induce sarcomas in animals. RSV is unusual among transforming viruses in that it retains intact genes for *gag*, *pol* and *env*. Usually the viral oncogene disrupts one or more of these three genes. This is commonly *gag* and the *onc* gene is expressed as a *gag–onc* fusion protein. Such viruses are defective and need to be propagated with a helper virus. The first step in viral replication is the reverse transcription of the viral RNA to produce linear double-stranded proviral DNA. In this process the termini of the RNA genome are duplicated to give terminal structures in the DNA known as long terminal repeats (LTRs). Circular proviral DNA is also found and is possibly the form which integrates into chromosomal DNA. The viral genome is expressed from the integrated proviral DNA which has promoter and polyadenylation signals in the LTRs. The integrated and unintegrated forms of the proviral DNA are colinear and so the transcript is identical to virion RNA and can therefore be packaged.

Several considerations suggest that the retroviruses will be good cloning vectors. In most of the cases so far examined, the viral *onc* genes have cellular homologues which are transcribed in normal cells. It seems likely therefore that the retroviruses are natural transducing agents, and therefore their genomes could be manipulated *in vitro* for use as vectors. Furthermore the viruses normally have a very wide host-range which can be extended by the process of pseudotyping. This is the phenomenon of one virus being packaged into the coat of another when the two viruses are grown in the same cell. Since the viral coat determines the host-range, it is possible to use such viral pseudotypes to introduce a virus into a cell which it would not normally be capable of infecting. Finally, retrovirus infection does not lead to cell death, but rather the infected or transformed cells grow and continue to produce infectious virions for many generations.

Several groups have now used retroviruses as vectors for the herpes simplex virus *tk* gene. Wei *et al.* [46] have inserted the *tk* gene into cloned proviral DNA of a deletion mutant of Harvey murine sarcoma virus (HaMuSV) such that the *onc* gene remains intact. The recombinant DNA was used to transform tk^- mouse cells. Foci of biologically transformed cells could then be picked and infected with a helper virus, in order to rescue the recombinant as a helper-dependent retrovirus. Similar experiments have been carried out with Moloney leukaemia virus (MLV) [47], and spleen necrosis virus (SNV) [48]. In the latter study it was observed that removal of sequences corresponding to the 3′ end of the *tk* mRNA increased the yield of recombinant virus by three orders of magnitude. This might reflect a requirement for the viral LTRs since the viral RNA has evolved to serve either as messenger or viral genome by having its promoter and polyadenylation sequences in the terminal repeats. The structure of a gene cloned into a retrovirus vector has therefore considerable importance. It will be necessary to exclude polyadenylation signals which would lead to truncated transcripts incapable of initiating the synthesis of + strands of viral DNA. Sorge and Hughes [49] have examined the fate of the intervening sequences in the human α-chorionic gonadotrophin gene inserted into the RSV genome to replace the *src* gene. After several generations of passage recombinant viruses appeared in which the human gene had lost intervening sequences, suggesting that spliced messengers can eventually be incorporated into virions. Such a mechanism would explain why cellular oncogenes incorporated into retrovirus genomes lack their intervening sequences.

The insertion of foreign DNA into the cloned proviral DNA of a retrovirus can produce a defective viral genome. It is therefore necessary to use helper viruses to propagate such recombinants just as we have seen for recombinants in SV40 or adenoviral vectors. A defective helper virus that is highly suited for this application has been described by Mann *et al.* [50]. They have deleted a 350 base pair sequence from the cloned proviral DNA of Moloney murine leukaemia virus (M-MuLV) to make a plasmid containing a defective virus that they call pMOVψ^-. The deleted DNA lies between the putative 5′ splice site of the *env* gene and the AUG initiation codon for the *gag* gene. It contains a sequence that is required in *cis* for the packaging of the viral RNA. Thus, although this proviral DNA contains all of the proteins specified by the virus, the viral RNA derived from it cannot be packaged. A defective virus such as one derived from a recombinant molecule constructed *in vitro* can be complemented by pMOVψ^- and packaged into virions as long as the recombinant genome contains the cis-acting packaging sequence. Cells that have had pMOVψ^- introduced into them will, for example, package the defective viral genome derived from a plasmid carrying the *E. coli gpt* gene within Maloney sarcoma virus (MSV) sequences (pMSVgpt). When these two plasmids are cotransfected into mouse cells they produce infective viral particles (Fig. 8.14). When these are used in a secondary infection of mouse cells they can be shown to express XGPRT (see Section 8.2.4 for details of the direct selection system for *gpt* activity).

The cloned helper provirus, pMOVψ^-, has been stably integrated into the chromosomes of a number of cell lines. One of these lines produced infectious virus presumably because of recombination between some DNA sequence in the host cell that can restore the packaging defect. Other lines would not make infectious virus although they would secrete 'empty' reverse transcriptase containing particles into the culture medium. One such line, ψ2, can be used as a host for cloned recombinant proviruses such as pMSVgpt and then produces high titres of infectious particles containing only the recombinant genome. This cell line is somewhat analogous to COS cells (Section 8.2) and 293 cells (Section 8.3) which provide helper functions for defective SV40 and adenovirus vectors respectively.

This ability to produce pure stocks of recombinant retroviruses is likely to be extremely useful. Recombinant viruses propagated in such cells could, for example, be used to introduce genes into the germ line of mice, as has already been demonstrated to be possible with murine leukaemia virus MuLV. When 32 cell embryos are infected with MuLV and reimplanted into foster mothers, up to half the mice that develop have proviral DNA integrated into chromosomes in some of their cells. These animals are mosaic for the presence of proviral DNA and this is not expressed. If, on the other

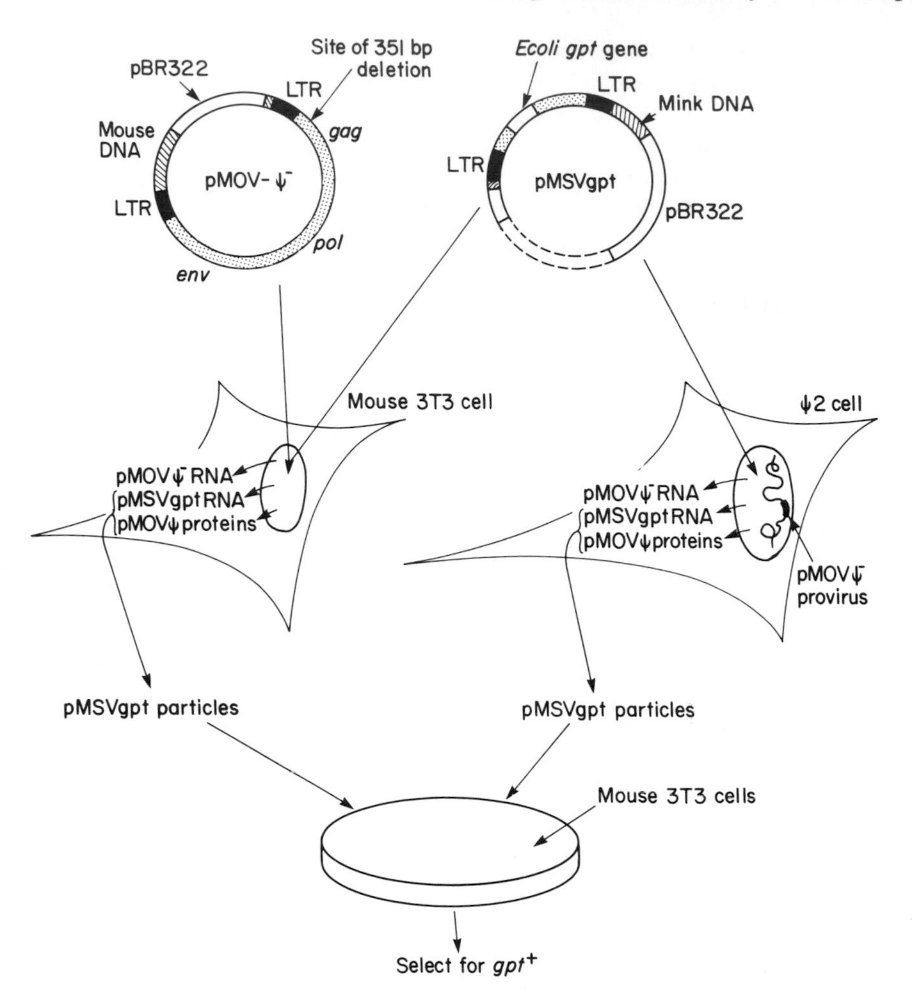

Figure 8.14 The production of pure stocks of recombinant retroviruses using the pMOV ψ helper or $\psi 2$ host cells.

hand, 8–10 day old embryos are infected *in utero* then virus is produced in all cells. Microinjection of MuLV proviral DNA has yet a different set of consequences. If injected into the nuclei of fertilized eggs, it integrates into the chromosomal DNA in multiple copies and is not expressed. If injected into the cytoplasm, however, single copies become integrated into the chromosomal DNA of all cells. About 10% of the mice that develop from these embryos become viremic and produce MuLV about three weeks after birth [51, 52]. The degree of expression of the integrated MuLV genome is apparently influenced by the site of chromosomal integration.

8.6 Vaccinia virus vectors

The viruses which we have considered so far have all been ones which replicate and are transcribed within the cell nucleus. Vaccinia virus on the other hand replicates and is transcribed in the cell cytoplasm. It cannot therefore rely on the host cell machinery for these functions and consequently its genome is large (180 kb). Against the problems of working with such a large DNA molecule is the impressive record of the use of the virus for vaccination against variola virus, the causative agent of smallpox. Vaccination against smallpox was first described almost 200 years ago [53] and is effective because the vaccinia and variola viruses are so closely related and because the subject is exposed to live vaccine. Recently foreign genes have been inserted into the vaccinia genome to produce viable recombinants. This opens the door to the possibility of producing hybrid viruses which could be used for the simultaneous immunization against several antigens.

The viral genome encodes about 200 polypeptides. About half of these genes are expressed early, that is to say before the onset of viral DNA replication. The early and late genes are interspersed throughout the genome and their pattern of expression is extremely complex. The viral DNA molecule has a large inverted terminal repetition of about 10 kb. At its ends this contains incompletely base paired loops which link the two DNA strands into one continuous chain. Next to this terminal region are two sets of 70 base pair repeats and the coding sequences for three polypeptides of 7.5, 19, and 42 kd that are synthesized immediately upon infection (Fig. 8.15). The mRNAs which encode these proteins are polyadenylated and have multiple 5′ ends which can be labelled with the nucleotide that initiates their biosynthesis indicating the absence of processing at the 5′ end [54]. These early mRNAs are not spliced. One transcript terminates near or within a specific hexanucleotide sequence that is tandemly repeated four times, and although there are AT-rich clusters near the termination sites, the specific AATAAA polyadenylation signal of nuclear mRNAs is not seen [55]. Similarly the sequences upstream from the position of the initiation of transcription share only limited homology with the eukaryotic or prokaryotic consensus sequences, and are comprised of a 60 base pair segment that is 80% AT-rich.

One easily characterized marker of early gene expression is the virally encoded thymidine kinase. The restriction fragment which carries this gene has been identified by translation assays (as described in Section 1.3.2(b)) and also by its ability to rescue infections with tk^- vaccinia virus. Since vaccinia virus DNA is not infectious, it is necessary first to infect cells with the tk^- virus and then attempt rescue by introducing restriction fragments by

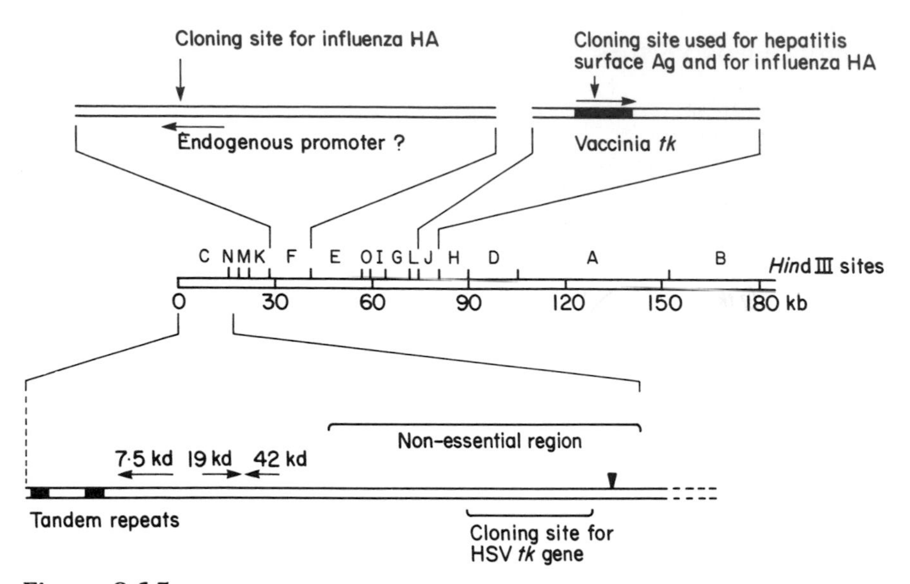

Figure 8.15 Physical map of the vaccinia genome showing sites at which foreign genes have been inserted and expressed.

coprecipitation with calcium phosphate (see Section 8.1) [56]. Recombination occurs *in vivo* to produce wild type virus. Conversely it is possible to make tk^- insertion mutants of vaccinia. A plasmid carrying the vaccinia *tk* gene interrupted by a segment of foreign DNA is introduced by coprecipitation with calcium phosphate into cells infected with wild type vaccinia virus. The *tk* gene inactivated by the insertion of foreign DNA recombines with the wild type vaccinia genome and tk^- recombinants can be selected.

An analogous approach has been used by workers in Moss's laboratory to introduce a functional herpes simplex virus *tk* gene into vaccinia. The HSV gene was inserted into a segment of vaccinia DNA known to be from a non-essential region. The recombinant was first constructed as a plasmid in such a way that the HSV *tk* gene was flanked by the non-essential vaccinia sequences. Upon introduction into vaccinia infected cells the plasmid can therefore recombine with the DNA of the wild type virus (Fig. 8.16). A vaccinia early promoter was also included upstream of the HSV *tk* gene in the plasmid. This was a 275 base pair fragment containing sequences for the initiation of transcription of the early gene for the 7.5 kd protein (see Fig. 8.15). The production of HSV thymidine kinase in cells infected with the resulting recombinant virus can be recognized by the unique ability of this enzyme to phosphorylate ^{125}I-labelled deoxycytidine [57].

This approach has been successfully extended to construct viable hybrid vaccinia viruses which contain the gene for the surface antigen of hepatitis B virus. Once again the foreign gene has been

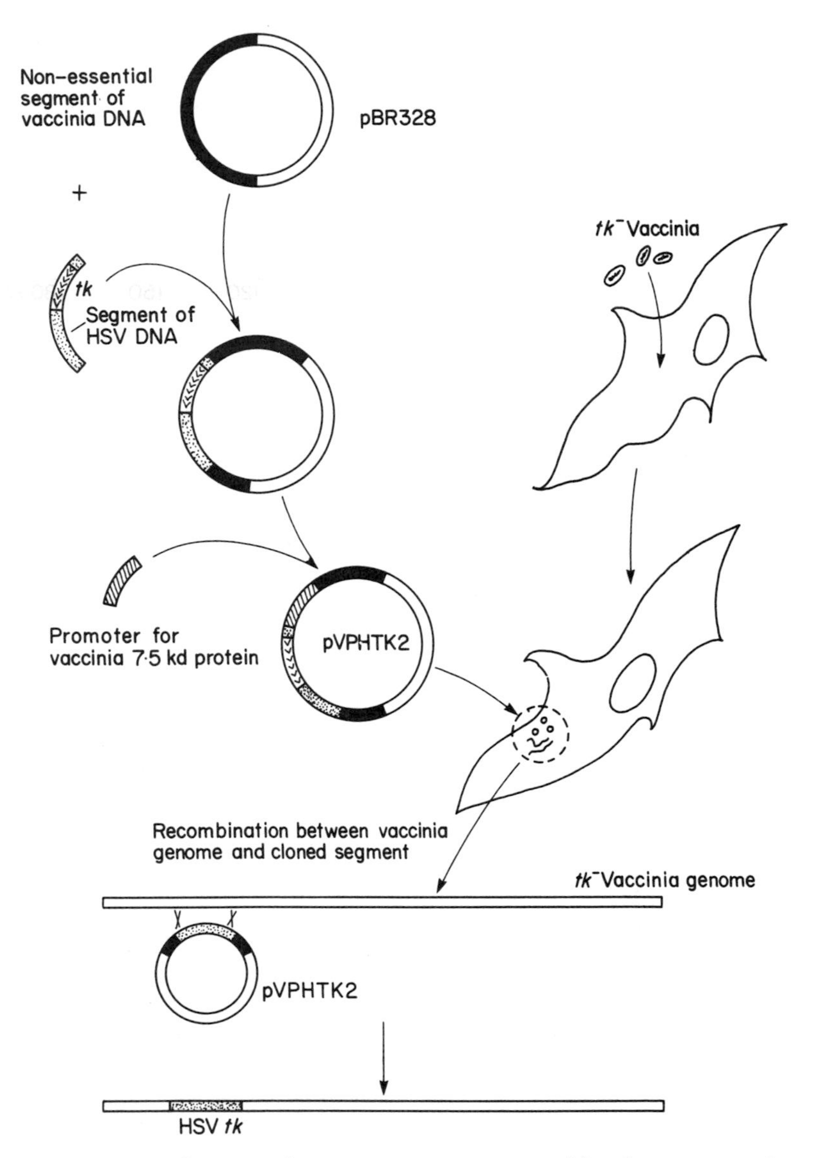

Figure 8.16 Construction of a recombinant vaccinia virus capable of expressing the thymidine kinase gene of herpes simplex virus.

placed under the control of the early vaccinia promoter for the 7.5 kd protein [58]. In this case the hepatitis B virus gene was inserted into the cloned *tk* gene of vaccinia. Recombination between the vaccinia genome and this plasmid effectively results in the insertion of the gene for the hepatitis surface antigen into the *tk* gene of vaccinia. Cells infected with this virus produce particles with similar appearance, sedimentation coefficient and antigenic properties to those produced by hepatoma cell lines. Furthermore innoculation of rabbits with the hybrid virus elicits the production of antibodies against the hepatitis surface antigen. A similar approach has been used to insert the gene for influenza virus haemagglutinin into the vaccinia *tk* gene. The innoculation of hamsters with this hybrid virus leads to the production of circulating antibodies which give protection against respiratory infection with influenza virus [59]. Another laboratory has used similar procedures to achieve expression of the influenza haemagglutinin [60]. In this case the gene has been inserted at a different non-essential region of the genome (see Fig. 8.15) and appears to be under the control of a vaccinia promoter at this site. The possibility therefore clearly exists for using this as a general approach for the production of combination vaccines. It also illustrates the significant progress which can be made having detailed knowledge of only a small proportion of a complex viral genome.

8.7 Maximizing expression

The characterization of regulatable promoters will allow their use within eukaryotic expression systems in a manner analogous to that described for prokaryotic promoters in Chapter 5. Upstream of nearly all eukaryotic nuclear genes, about 30 nucleotides from the transcription initiation site, there is a variant on the sequence TATA(A/T)A(A/T) which seems to act as a signal to position the 5′ ends of RNAs. Additional regulatory elements are now being recognized. The early promoter of SV40 has a 72 base pair tandem repeat further upstream, one copy of which is absolutely required for the expression of early genes. When the 72 base pair repeat is linked to the rabbit β-globin gene, then transcription of the globin gene is greatly enhanced in cultured human cells [61]. The effect is cis-acting and can be propagated over several kilobases; it is independent of the orientation of the 72 base pair repeat and of the position of the sequence with respect to the gene. Analogous enhancer elements have been found in the long terminal repeats (LTRs) of Maloney sarcoma virus (MSV) and Rous sarcoma virus (RSV). The LTR of MSV contains a 73 base pair repeat which can substitute for the 72 base pair repeat in the SV40 genome [62]. The effect of mutagenesis upon the SV40 72 base pair repeat has been studied in a construct in which it directs the expression of the prokaryotic gene for chloramphenicol transacetylase. These studies have identified critical nucleotides within a short sequence having a degree of homology with segments of other enhancer sequences.

The LTR of mouse mammary tumour virus (MMTV) has an additional advantage in that it contains sequences which mediate a transcriptional response to glucocorticoids. The transcription of the virus can be regulated by dexamethasone and it has been shown that exogenous genes fused to the MMTV LTR can also be hormonally regulated [63]. In these experiments the MMTV promoter was linked to the cDNA of the mouse dihydrofolate reductase gene and the hormonal response of its transcription was demonstrated both by measuring the sensitivity of transformed cells to methotrexate and by directly assaying the amount of DHFR protein produced.

Another inducible promoter is that of the metallothionein gene. Metallothioneins are metal binding proteins which function in heavy metal detoxification. The transcription of the mouse metallothionein I gene is induced by heavy metals. The cloned mouse metallothionein I gene linked to the *E. coli gpt* gene has been introduced into human *hgprt*$^-$ cells using the direct transformation procedures discussed earlier in this chapter. The transfected genes retain their ability to be induced by cadmium in this new environment [64]. In the same study, the regulatory region from the metallothionein I gene was fused to the HSV thymidine kinase gene. The *tk* gene then becomes regulated by cadmium when introduced into mouse cells. The sequences necessary for this regulation lie within 148 base pairs of the site for the initiation of transcription. The regulation is maintained when the gene is incorporated into an SV40 derived vector and introduced into monkey cells. Since the metallothionein gene does not undergo tissue specific expression, its promoter can probably be used as a regulatory element in a variety of cloning systems.

Heat shock promoters may provide another set of useful inducible regulatory elements. The heat shock response was originally extensively studied in *Drosophila* but has since been shown to be a ubiquitous, cellular response. When *Drosophila* cells are shifted from 25–37° C, the pre-existing polysomes break down and the transcription of some eight different heat shock genes is initiated leading to a dramatic alteration in the pattern of protein synthesis. A similar response is seen in eukaryotes as diverse as yeast and the mammals and the major heat shock proteins appear to be evolutionarily conserved. The response is mediated by a conserved cis-acting element. The *Drosophila* genes for example remain thermo-inducible when injected into *Xenopus* oocytes [65] or introduced into COS cells using a vector containing an SV40 replication origin [66]. This latter system has been used to analyse deletion mutants upstream of the transcription initiation site for the major heat shock protein of *Drosophila*. A functional element has been identified between residues −47 and −66. When sequences within this region are examined from other heat shock loci, it has proved possible to identify a 15 nucleotide consensus sequence. When sequences similar to the consensus are chemically synthesized and placed upstream of the HSV thymidine kinase gene, then heat inducible expression of thymidine kinase can be observed when the construct is introduced either into monkey cells or *Xenopus* oocytes [67].

The strength of a promoter can be gauged by including it in an expression vector so that it directs the synthesis of a gene product that can be readily assayed. This has been done using the *E. coli gpt* gene or *gal* gene, that express guanine phosphoribosyl transferase and galactose kinase respectively. The bacterial gene for chloramphenicol acetyl transferase (CAT) cloned into pSV2 has found extensive use for this purpose. The enzyme can be conveniently assayed by its ability to acetylate ^{14}C-chloramphenicol and there is no evidence of any endogenous chloramphenicol modifying enzyme in any mammalian tissue that has been examined. Amongst the several promoters that have been assayed using this system, the most effective was the LTR of Rous sarcoma virus which promotes a high level of CAT expression in a variety of cell types [68]. As well as measuring promoter function, these assays can also be used to determine the optimum conditions for transforming cells. The transient expression assay is made on the order of 48 hours after the introduction of the recombinant DNA into the cells. There seems to be, however, a direct correlation between the level of transient expression and the proportion of cells that go on to become stably transformed. The system is therefore finding considerable use in defining the conditions for the transformation of cell lines which have previously given the impression of being difficult to transform [69].

8.8 Whole animal transformation

The regulatory promoters described above will find their way into expression vectors. Attempts to reintroduce cloned genes into animal cells are no longer restricted to experiments carried out with cultured cells. Experiments with the fruit fly *Drosophila melanogaster*, and with mice have shown that cloned genes can be reintroduced into the chromosomes of these organisms so that they are inherited in a Mendelian manner.

8.8.1 *Drosophila* transformation

The transformation system developed for *Drosophila* utilizes a transposon known as the P-factor. This is the causative agent of one of two genetically distinct types of hybrid dysgenesis. Hybrid dysgenesis occurs in crosses between certain strains of flies leading to lowered fertility, recombination in males, and increased frequencies of mutations. In the P–M system dysgenesis is seen in crosses between M (maternal) strain females and P (paternal) strain males. Reciprocal crosses give normal progeny. This can be rationalized by supposing that the transposition of P-elements is regulated by a repressor. Such a molecule would be absent from the eggs of M-females. When such eggs are fertilized by sperm from P-males, the paternal chromosomes enter an environment in which the transposition of their P-elements is no longer repressed. The I–R system is similar in that crosses of I (inducer) strain males and R (reactive) strain females result in dysgenesis.

The cloning of the *white* gene (see Section 1.3.3) allowed Rubin and his colleagues to investigate the structure of several mutations at this allele induced by P–M dysgenesis. These appear to be caused by the insertion of defective P-elements which are too short to be fully functional P-factors. These P-elements have been used to screen a library of phages constructed from a P-strain and full length P-factors have been isolated [70]. The P-factor is 2907 base pairs long and has inverted terminal repeats of 31 base pairs. There are three open reading frames that would encode polypeptides of 238, 264, and 218 amino-acids. P-factors and P-elements, like other transposable elements are flanked by a duplication of an oligonucleotide sequence (in this case eight base pairs long) at the insertion site. The insertion sites appear not to be entirely random since three of the five dysgenic mutations of *white* have insertions at the same site. Chromosome rearrangements also occur at high frequency in the germline of dysgenic individuals at particular hotspots.

Spradling and Rubin found that by microinjecting P-factor clones into embryos of M-cytotype they could achieve the same effects as in a dysgenic cross. In their experiments they used a recipient strain that has a mutation *singed*^{weak} (*sn*^{w}) due to the presence of a defective P-element at the *sn* locus. Hybrid dysgenesis results either in a more extreme mutation or the reversion of this allele to give the wild phenotype. This effect could be scored in the progeny of injected embryos. In the process the elements of the bacterial vector are not incorporated into the fly chromosomes but rather the P-factor itself transposes from the plasmid vector into the chromosomes.

Defective P-elements can be used as vectors to reintroduce genes into *Drosophila*. This was first shown by Spradling and Rubin who introduced the *Drosophila* gene for xanthine dehydrogenase into a P-element such that it was flanked by the 31 nucleotide sequence repeat at the P-element terminus [71]. This was microinjected together with an intact 'helper' P-element into embryos deficient for this gene. Such embryos would normally give rise to flies with 'rosy' coloured eyes. Mosaic individuals are produced in the first generation as there is no way of ensuring that the injected DNA goes into a particular cell type. Stable transformants can, however, be recognized in subsequent generations by their wild type eye colour (Fig. 8.17). The injected genes transpose into new loci in the chromosomes of the host but retain their tissue-specific and developmental pattern of regulation. This has now also been shown to be true for several other *Drosophila* genes. A proportion of the injected embryos gives rise to flies in which the wild type eye colour is stably inherited. This is a result of the chromosomal integration of the P-element carrying the xanthine dehydrogenase gene. Integration does not occur by homologous recombination with endogenous sequences, but is directed by the transposon to random sites. A number of *Drosophila* genes have been reintroduced into flies using this approach. In the cases studied so far, correct tissue-specific and

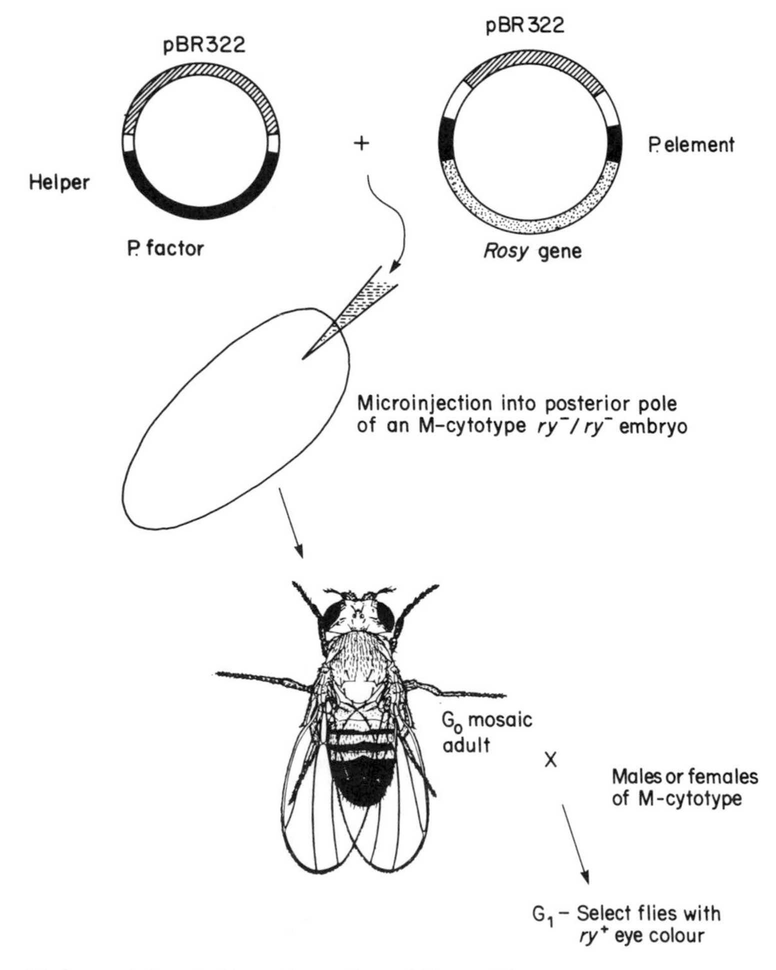

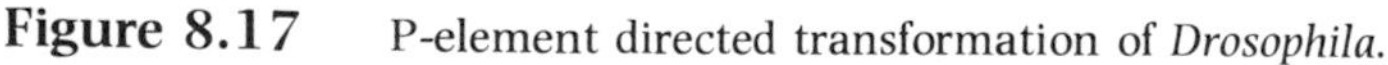
Figure 8.17 P-element directed transformation of *Drosophila*.

developmental expression are retained irrespective of the new chromosomal site occupied by a gene and its flanking sequences. This will permit a detailed study of sequences which act in 'cis' to regulate gene expression. The system will clearly be invaluable for the study of *cis* acting regulatory sequences that control correct developmental gene expression.

8.8.2 Mouse transformation

There are several possible routes to introduce foreign genes into a mouse. One approach would be to introduce the gene into cultured teratocarcinoma cells using one of the techniques described earlier in this chapter. Teratocarcinoma cells can be injected into the pre-implanted mouse embryo when it is at the blastocyst stage. Several laboratories have shown that the teratocarcinoma cells participate in development and if they carry a suitable genetic marker, then the production of a genetically mosaic organism can be readily seen [72]. This combined approach has yet to be tried.

A more direct approach has already had demonstrable success. This involves injecting recombinant DNA molecules into the pronuclei of fertilized eggs, followed by the implantation of the eggs into the reproductive tracts of foster mothers. It was first demonstrated that SV40 DNA injected in this way gave rise to mice with the viral sequences in their tissues [73]. Subsequently a recombinant plasmid containing SV40, HSV and pBR322 sequences was injected and shown to be present in some of the adult mice. In this experiment expression of the HSV thymidine kinase gene could not be detected [74]. In some cases, however, it became clear that the acquired genes were expressed in the adult mice and sometimes in their offspring. This was shown for the rabbit β-globin gene, the human β-globin gene, as well as the HSV thymidine kinase gene. In one study the HSV *tk* structural gene was joined to the regulatory region of the mouse metallothionein I gene [75]. From 69 fertilized eggs which were injected with this recombinant plasmid, ten adult mice developed which carried the exogenous genes. Of these, seven expressed the viral thymidine kinase and moreover the enzyme was inducible by giving cadmium to the mice.

The tissue-specificity of expression seems to vary between experiments. Cases have been reported in which globin genes are not transcribed at all, or are expressed in muscle rather than erythroid tissue. The expression of the herpes *tk* gene under the metallothionein promoter shows great specificity, the highest level of expression being seen in the liver and kidney. This pattern follows the expression of the endogenous metallothionein gene. The most striking example of tissue-specific expression has been from functional immunoglobulin genes injected into mouse embryos. The gene was for a κ light chain cloned from myeloma cells in which the gene had undergone rearrangement (see Section 6.1.3(c)). These 'transgenic mice' showed expression of the exogenous immunoglobulin gene in their spleens (presumably in B lymphocytes), but not in their livers.

At the moment it would seem that there are no clear rules governing the probability of attaining expression of genes introduced into mice in this way. The integration of genes is not chromosome specific and therefore the chromosome location assumed by the injected genes could well influence their expression. The injected genes can lead to mutation when inserted at certain chromosomal positions and examples of homozygous lethal mutations resulting from such insertions have been reported [76]. Nevertheless the system offers enormous potential.

This is illustrated by other experiments carried out by Palmiter, Brinster and their coworkers who injected mouse embryos with a

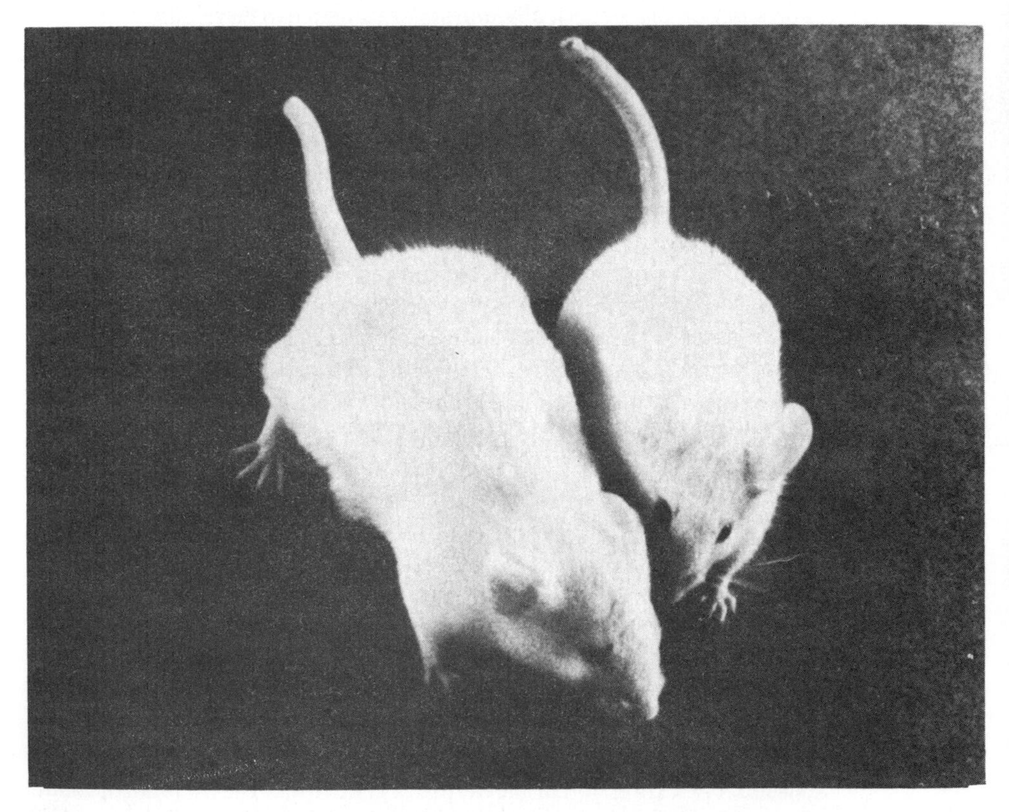

Figure 8.18 Increase in body weight of a mouse transformed by multiple copies of the growth hormone gene. The transformed mouse is shown alongside a non-transformed sibling (photograph courtesy of R. Brinster).

DNA fragment containing the rat growth hormone gene fused to the regulatory region of the mouse metallothionein I gene [77]. These experiments used linear DNA fragments rather than plasmids since these integrate more efficiently into the mouse chromosomes. Of the 21 adult mice which developed, seven contained the fusion gene of which six grew significantly larger than their littermates. An example of an animal with twice the body weight of its littermate is shown in Fig. 8.18. The growth hormone levels in three of the mice were between 200–800 times greater than in the controls and this reflected the elevated level of growth hormone mRNA that could be detected in the livers. There was also a broad correlation with the number of growth hormone genes per cell, the mouse having acquired the greatest number of genes (35 per cell), having the highest level of growth hormone. The effect of heavy metals on the regulation of these hybrid genes is as yet uncertain. This is perhaps one of the more efficient expression systems that has so far been described. The whole animal is certainly the most sophisticated and efficient cell culture system that we know. Vastly elevated levels of the hormone are produced and there are none of the problems of post-translational modification of the product which are encountered with the expression vectors described in Chapter 5. Many hormones are currently prepared from animal tissues. It is not inconceivable that in the future we could see the breeding of livestock containing multiple copies of pharmaceutically useful peptides within their genomes.

References

1. Szybalski, E. H. and Szybalska, W. (1962) Genetics of human cell lines. DNA mediated heritable transformation of a biochemical trait. *Proc. Natn Acad. Sci. USA*, **48**, 2026–34.
2. Graham, F. L. and Van Der Eb, A. J. (1973) A new techniques for the assay of infectivity of human adenovirus 5 DNA. *Virology*, **52**, 456–67.
3. Wigler, M., Pellicer, A., Silverstein, S. *et al.* (1979) DNA mediated transfer of the adenine phosphoribosyl transferase locus into mammalian cells. *Proc. Natn Acad. USA*, **76**, 1373–6.
4. Wigler, M., Sweet, R., Sim, G.-K. *et al.* (1979) Transformation of mammalian cells with genes from prokaryotes and eukaryotes. *Cell*, **16**, 777–85.
5. Perucho, M., Hanahan, D. and Wigler, M. (1980) Genetic and physical linkage of exogenous sequences in transformed cells. *Cell*, **22**, 309–18.
6. Huttner, K. M., Barbosa, J. A., Scangos, G., Pravtchera, D. and Ruddle, F. H. (1981) DNA mediated gene transfer without carrier DNA. *J. Cell Biol.*, **91**, 153–6.
7. Perucho, M., Hanahan, D., Lipsich, L. and Wigler, M. (1980) Isolation of

the chicken thymidine kinase gene by plasmid rescue. *Nature*, **285**, 207–10.

8. Lowy, I., Pellicer, A., Jackson, J. F. *et al.* (1980) Isolation of transforming DNA: cloning the hamster APRT gene. *Cell*, **22**, 817–23.
9. Goldfarb, M. P., Shimizu, K., Perucho, M. and Wigler, M. H. (1982) Isolation and preliminary characterisation of a human transforming gene from T24 bladder carcinoma cells. *Nature*, **296**, 404–9.
10. Shimizu, K., Goldfarb, M., Perucho, M. and Wigler, M. (1983) Isolation and preliminary characterisation of the transforming gene of a human neuroblastoma cell line. *Proc. Natn Acad. Sci. USA*, **80**, 383–7.
11. Tooze, J. (ed.) (1981) *DNA Tumour Viruses*, Cold Spring Harbor Laboratory, New York.
12. Goff, S. P. and Berg, P. (1976) Construction of hybrid viruses containing SV40 and lambda phage DNA segments and their propagation in cultured monkey cells. *Cell*, **9**, 695–705.
13. Mulligan, R. C., Howard, B. H. and Berg, P. (1979) Synthesis of rabbit β-globin in cultured monkey kidney cells following infection with a SV40 β-globin recombinant genome. *Nature*, **277**, 108–14.
14. Hamer, D. H. and Leder, P. (1979) Splicing and the formation of stable RNA. *Cell*, **18**, 1299–302.
15. Gruss, P., Lai, C.-J., Dhar, R. and Khoury, G. (1979) Splicing as a requirement for biogenesis of functional 16S mRNA of Simian virus 40. *Proc. Natn Acad. Sci. USA*, **76**, 4317–21.
16. Gruss, P. and Khoury, G. (1980) Rescue of a splicing defective mutant by insertion of an heterologous intron. *Nature*, **286**, 634–7.
17. Gruss, P., Efstratiadis, A., Karathanasis, S., Konig, M. and Khoury, G. (1981) Synthesis of stable unspliced mRNA from an intronless Simian virus 40 – rat preproinsulin gene recombinant. *Proc. Natn Acad. Sci. USA*, **78**, 6091–5.
18. Gething, M. J. and Sambrook, J. (1981) Construction of influenza haemagglutinin genes that code for intracellular and secreted forms of the protein. *Nature*, **300**, 598–603.
19. Gluzman, Y. (1981) SV40 transformed Simian cells support the replication of early SV40 mutants. *Cell*, **23**, 175–82.
20. Myers, R. M. and Tjian, R. (1980) Construction and analysis of Simian virus 40 origins defective in tumour antigen binding and DNA replication. *Proc. Natn Acad. Sci. USA*, **77**, 6491–5.
21. Lusky, M. and Botchan, M. (1981) Inhibition of SV40 replication in Simian cells by specific pBR322 DNA sequences. *Nature*, **293**, 79–81.
22. Mellon, P., Parker, V., Gluzman, Y. and Maniatis, T. (1981) Identification of DNA sequences required for transcription of the human α1 globin gene in a new SV40 host vector system. *Cell*, **27**, 279–88.
23. Berg, P. (1981) Dissection and reconstruction of genes and chromosomes. *Science*, **213**, 296–303.
24. Breathnach, R. and Harris, B. A. (1983) Plasmids for the cloning and expression of full length stranded cDNAs under control of the SV40 early or late promoter. *Nucl. Acids Res.*, **11**, 7119–36.
25. Mulligan, R. C. and Berg, P. (1980) Expression of a bacterial gene in mammalian cells. *Science*, **209**, 1422–7.

26. Mulligan, R. C. and Berg, P. (1981) Seclection for animal cells that express the *Escherichia coli* gene coding for xanthine–guanine-phosphoribosyltransferase. *Proc. Natn Acad. Sci. USA*, **78**, 2072–6.
27. Mulligan, R. C. and Berg, P. (1981) Factors governing the expression of a bacterial gene in mammalian cells. *Mol. Cell Biol.*, **1**, 449–59.
28. Colbere-Garapin, F., Horodniceanu, F., Kourilsky, P. and Garapin, A.-C. (1981) A new dominant hybrid selective marker for higher eukaryotic cells. *J. Mol. Biol.*, **150**, 1–14.
29. Canaani, D. and Berg, P. (1982) Regulated expression of human interferon β^1 gene after transduction into cultured mouse and rabbit cells. *Proc. Natn Acad. Sci. USA*, **79**, 5166–70.
30. Subramani, S., Mulligan, R. and Berg, P. (1981) Expression of the mouse dihydrofolate reductase complementary deoxyribonucleic acid in Simian virus 40 vectors. *Mol. Cell Biol.*, **1**, 854–64.
31. Kaufman, R. J. and Sharp, P. A. (1982) Amplification and expression of sequences contransfected with a modular dihydrofolate reductase complementary DNA gene. *J. Mol. Biol.*, **159**, 601–22.
32. Jones, N. and Shenk, T. (1978) Isolation of deletion and substitution mutants of adenovirus type 5. *Cell*, **13**, 181–8.
33. Jones, N. and Shenk, T. (1979) Isolation of adenovirus type 5 host range deletion mutants defective for transformation of rat embryo cells. *Cell*, **17**, 683–9.
34. Stow, N. (1982) The infectivity of adenovirus genomes lacking DNA sequences from their left hand termini. *Nucl. Acids Res.*, **10**, 5105–20.
35. Gluzman, Y., Reichl, H. and Solnick, D. (1982) in *Eukaryotic Viral Vectors* (ed. Y. Gluzman), Cold Spring Harbor Laboratory, New York, p. 187.
36. Thummel, C., Tjian, R. and Grodzicker, T. (1981) Expression of SV40 T antigen under control of adenovirus promoters. *Cell*, **23**, 825–36.
37. Solnick, D. (1981) Construction of an adenovirus – SV40 recombinant producing SV40 T antigen from an adenovirus late promoter. *Cell*, **24**, 135–43.
38. Law, M.-F., Lowy, D. R., Dovoretzky, I. and Howley, P. M. (1981) Mouse cells transformed by bovine papilloma virus contain only extra-chromosomal viral DNA sequences. *Proc. Natn Acad. Sci. USA*, **78**, 2727–31.
39. Moar, M. H., Campo, M. S., Laird, H. M. and Jarrett, W. F. M. (1981) Unintegrated viral DNA sequences in a hamster tumour induced by bovine papilloma virus. *J. Virol.*, **39**, 945–9.
40. Lowy, D. R., Dovoretzky, I., Shober, R. *et al.* (1980) *In vitro* tumorigenic transformation by a defined subgenomic fragment of bovine papilloma virus DNA. *Nature*, **287**, 72–4.
41. Sarver, N., Gruss, P., Law, M.-F., Khoury, G. and Howley, P. M. (1981) Bovine papilloma virus deoxyribonucleic acid – a novel eukaryotic cloning vector. *Mol. Cell Biol.*, **1**, 486–96.
42. Pavlakis, G. N. and Hamer, D. H. (1983) Regulation of a metallothionein – growth hormone hybrid gene in bovine papilloma virus. *Proc. Natn Acad. Sci. USA*, **80**, 397–401.
43. DiMaio, D., Treisman, R. and Maniatis, T. (1982) Bovine papilloma virus vector that propagates as a plasmid in both mouse and bacterial cells. *Proc. Natn Acad. Sci. USA*, **79**, 4030–4.

44. Zinn, K., DiMaio, D. and Maniatis, T. (1983) Identification of two distinct regulatory regions adjacent to the human β interferon gene. *Cell*, **34**, 865–79.
45. Weiss, R. A., Teich, N. M., Varmus, H. and Coffin, J. M. (1982) in *RNA Tumour Viruses*. 2nd edn (eds. R. A. Weiss *et al.*), Cold Spring Harbor Laboratory, New York.
46. Wei, C.-M., Gibson, M., Spear, P. G. and Scolnick, E. M. (1981) Construction and isolation of a transmissible retrovirus containing the *src* gene of Harvey murine sarcoma virus and the thymidine kinase gene of herpes simplex virus type I. *J. Virol.*, **39**, 935–44.
47. Tabin, C. J., Hoffmann, J. W., Goff, S. P. and Weinberg, R. A. (1982) Adaptation of a retrovirus as a eukaryotic vector transmitting the herpes simplex virus thymidine kinase gene. *Mol. Cell Biol.*, **2**, 426–36.
48. Shimotohno, K. and Temin, H. M. (1981) Formation of infectious progeny virus after insertion of herpes simplex thymidine kinase gene into DNA of an avian retrovirus. *Cell*, **26**, 67–77.
49. Sorge, J. and Hughes, S. H. (1982) Splicing of intervening sequences introduced into an infectious retroviral vector. *J. Mol., Appl. Genet.*, **1**, 547–59.
50. Mann, R., Mulligan, R. C. and Baltimore, D. (1983) Construction of a retrovirus packaging mutant and its use to produce help free defective retrovirus. *Cell*, **33**, 153–9.
51. Jaenisch, R. (1979) Moloney leukaemia virus gene expression and gene amplification in preleukaemic and leukaemic Balb/Mo mice. *Virology*, **93**, 80–90.
52. Harbers, K., Jahner, D. and Jaenisch, R. (1981) Microinjection of cloned retroviral genomes into mouse zygotes: integration and expression in the animal. *Nature*, **293**, 540–2.
53. Jenner, E. (1798) An enquiry into the causes and effects of the variolae vacciniae, a disease discovered in some western counties of England, particularly Gloucestershire and known by the name of cow pox (Reprint: Cassell, London, 1896).
54. Venkatesan, S., Baroudy, B. M. and Moss, B. (1981) Distinctive nucleotide sequences adjacent to multiple initiation and termination site of an early vaccinia virus gene. *Cell*, **25**, 805–13.
55. Venkatesan, S., Gershowitz, A. and Moss, B. (1982) Complete nucleotide sequences of two adjacent early vaccinia virus genes located within the inverted terminal repetition. *J. Virol.*, **44**, 637–46.
56. Weir, J. P., Bajszar, G. and Moss, B. (1982) Mapping of the vaccinia virus thymidine kinase gene by marker rescue and by cell free translation of selected mRNA. *Proc. Natn Acad. Sci. USA*, **79**, 1210–4.
57. Mackett, M., Smith, G. L. and Moss, B. (1982) Vaccinia virus: a selectable eukaryotic cloning and expression vector. *Proc. Natn Acad. Sci. USA*, **79**, 7415–9.
58. Smith, G. L., Mackett, M. and Moss, B. (1983) Infectious vaccinia virus recombinants that express hepatitis B virus surface antigen. *Nature*, **302**, 490–5.
59. Smith, G. L., Murphy, B. R. and Moss, B. (1983) Construction and characterisation of an infectious vaccinia virus recombinant that expresses the influenza haemagglutinin gene and induces resistance to

influenza virus infection in hamsters. *Proc. Natn Acad. Sci. USA*, **80**, 7155–9.

60. Panicali, D., Davis, S. W., Weinberg, R. L. and Paoletti, E. (1983) Construction of live vaccines by using genetically engineered pox viruses biological activity of recombinant vaccinia virus expressing influenza virus haemagglutinin. *Proc. Natn Acad. Sci. USA*, **80**, 5364–8.
61. Banerji, T., Rusconi, S. and Schaffner, W. (1981) Expression of a β-globin gene is enhanced by remote SV40 DNA sequences. *Cell*, **27**, 299–308.
62. Weiher, H., Konig, M. and Gruss, P. (1983) Multiple point mutations affecting the Simian virus 40 enhancer. *Science*, **219**, 626–31.
63. Lee, F., Mulligan, R., Berg, P. and Ringold, G. (1981) Glucocorticoids regulate expression of dihydrofolate reductase cDNA in mouse mammary tumour virus chimaeric plasmids. *Nature*, **294**, 228–32.
64. May, K. E., Warren, R. and Palmiter, R. D. (1982) The mouse metallothionin 1 gene is transcriptionally regulated by cadmium following transfection into human or mouse cells. *Cell*, **29**, 99–108.
65. Bienz, M. and Pelham, H. R. B. (1982) Expression of a *Drosophila* heat shock protein in *Xenopus* oocytes: conserved and convergent regulatory signals. *EMBO J.*, **1**, 1583–8.
66. Pelham, H. R. B. (1982) A regulatory upstream promoter element in the *Drosophila* HSP70 heat shock gene. *Cell*, **30**, 517–28.
67. Pelham, H. R. B. and Bienz, M. (1982) A synthetic heat shock promoter element confers heat inducibility on the herpes simplex virus thymidine kinase gene. *EMBO J.*, **1**, 1473–7.
68. Gorman, C., Merlino, G. T., Willingham, M. C., Pustan, I. and Howard, B. H. (1982) The Rous Sarcoma virus long terminal repeat is a strong promoter when introduced into a variety of eukaryotic cells by DNA mediated transfection. *Proc. Natn Acad. Sci. USA*, **79**, 6777–81.
69. Gorman, C., Padmanabhan, R. and Howard, B. H. (1983) High efficiency DNA mediated transformation of primate cells. *Science*, **221**, 551–3.
70. O'Hare, K. and Rubin, G. (1983) Structures of P transposable elements and their sites of insertion and excision in the *Drosophila melanogaster* genome. *Cell*, **34**, 24–35.
71. Rubin, G. and Spradling, A. (1982) Genetic transformation of *Drosophila* with transposable element vectors. *Science*, **218**, 348–53.
72. Mintz, B. and Illmensee, K. (1975) Normal genetically mosaic mice produced from malignant teratocarcinoma cells. *Proc. Natn Acad. Sci. USA*, **72**, 3585–9.
73. Jaenisch, R. and Mintz, B. (1974) Simian virus 40 DNA sequences in DNA of healthy adult mice derived from preimplantation blastocysts injected with viral DNA. *Proc. Natn Acad. Sci. USA*, **71**, 1250–4.
74. Gordon, J. N., Scangos, G. A., Plotkin, D. J. Barbosa, J. A. and Ruddle, F. H. (1980) Genetic transformation of mouse embryos by microinjection of purified DNA. *Proc. Natn Acad. Sci. USA*, **77**, 7380–4.
75. Palmiter, R. D., Chen, H. Y. and Brinster, R. L. (1982) Differential regulation of metallothionin – thymidine kinase fusion genes in transgenic mice and their offspring. *Cell*, **29**, 701–10.

76. Wagner, E., Covarrubias, L., Stewart, T. A. and Mintz, B. (1983) Prenatal lethalities in mice homozygous for human growth hormone gene sequences integrated in the germ line. *Cell*, **35**, 647–55.
77. Palmiter, R. D., Brinster, R. L., Hammer, R. E. *et al.* (1982) Dramatic growth of mice that develop from eggs microinjected with metallothionin growth hormone fusion genes. *Nature*, **300**, 611–5.

Index